Handbook of SCIENCE COMMUNICATION

compiled by

Anthony Wilson
formerly of the Science Museum, London

with contributions from

Jane Gregory
University College, London
Steve Miller
University College, London
and
Shirley Earl
Napier University, Edinburgh
formerly of the Robert Gordon University, Aberdeen

This book is the result of an initiative of the Education Department of the Institute of Physics

Institute of Physics Publishing
Bristol and Philadelphia

British Library Cataloguing-in-Publication Data

A catalogue record for this book is available from the British Library.

ISBN 0 7503 0518 5 (pbk)

Library of Congress Cataloging-in-Publication Data are available

First printed 1998
Reprinted 1999

Published by Institute of Physics Publishing, wholly owned by The Institute of Physics, London

Institute of Physics Publishing, Dirac House, Temple Back, Bristol BS1 6BE, UK

US Office: Institute of Physics Publishing, The Public Ledger Building, Suite 1035, 150 South Independence Mall West, Philadelphia, PA 19106, USA

Typeset in PageMaker 6.5 by Anthony Wilson
Printed in the UK by J W Arrowsmith Ltd, Bristol

CONTENTS

EDITOR'S NOTE

The *Handbook of Science Communication* is intended for use in higher education science departments and is appropriate for Science Communications courses.

It is based on an initiative set up by the Education Department of the Institute of Physics to put together a pack suitable for use in universities and colleges.

The aim of the Education Department in putting the original material forward for publication was to provide a tool that all science teaching staff could use to help students speak and write effectively. It addresses the lack of a book for physicists and other scientists specifically on core communication/presentation skills.

The material in this book is in two parts. The first part, 'Practical Science Communication', is a basic collection of practical advice and guidance for students. The second part, 'Communication Theory and Practice', is an introduction to the academic study of communication and includes some more detailed discussion of communication theories.

Kathryn Cantley
August 1998

FOREWORD

> 'We know you're brilliant academically, but what else can you do?'

The voice of the employer is being increasingly heard, particularly in universities where choices for future employment become increasingly complex.

The Confederation of British Industry has emphasised the importance of learning certain basic skills within education. The same message has come from the Royal Society of Arts. In his reviews on education in the UK Sir Ron Dearing reported that employers were emphasising more and more the need to build up competence in the key skills of communication, numeracy and information technology:

> 'There is much concern in all quarters about current standards of achievement in communication ... No other issue attracted more comment from employers and higher education'—*Sir Ron Dearing*

It is therefore particularly appropriate that the Institute of Physics has supported publication of this *Handbook of Science Communication*. There is a need for it.

It is curious that students, who may have sat through hours of lectures, many of varying quality, have not learned from the experience something of the art of communication: how to speak up, how to hold the attention and interest of the audience, how to vary intonation, the value of pauses, how to use body language, and how to make effective use of visual aids.

Many a talk at a conference is marred by speakers who think only of the content of their material without considering the importance of their presentation. We can all recall the speaker who never raises their eyes to look at those they are addressing, and of visual aids only legible from the first two rows.

It is equally curious that when answering examination questions students too often forget that they are writing for someone to read what they have written. Those hurriedly written, turgid, and almost illegible lines of continuous prose through which examiners have to struggle can seldom put them into a favourable frame of mind.

Something more carefully written and structured would more obviously reveal students' knowledge and ability. There is much wisdom on Written Communication in chapter 3 of this Handbook: let us hope that many students will read it.

A spokesman from the UK Association of Graduate Recruiters was reported recently as saying how large is the number of graduates who do not know how to write a covering letter when submitting a CV, even though their future career could depend on it. Statements like this show how easy it is to assume that these are things that someone else teaches, or are taught at some other level of education, when quite often they are not.

There are other things which may be taken for granted. We tend to assume, often incorrectly, that students know how to gather information and how to use libraries. This is part of another highly relevant skill: learning how to learn. But what attention is ever given at the tertiary level to such a basic skill as learning how to take notes? This is not a trivial skill (as many of us will recollect) and perhaps more attention might be given to that in future. Listening, note-taking, and writing reports are important parts of the communication process.

Sir Ron Dearing in his review also stated that:

> '... employers in particular have stressed the importance of developing wider skills, including inter-personal skills, particularly team-working, presentational skills, a problem-solving approach, and the ability to manage one's own learning.'

It is therefore particularly encouraging that one chapter is devoted to Working in Groups and Teams. It considers the importance of these skills, which too many of us assume are obvious and that everyone possesses, when this is certainly not the case.

The material in this book is relevant to undergraduate and graduate students and it is to be hoped that they will read it. Let us hope that more and more departments will come to appreciate the value of incorporating science communication in their courses. It would greatly improve the quality of the learning process and meet the expectations which industry and other employers have of those entering upon their careers.

John L Lewis
The Institute of Physics
August 1998

Part 1
PRACTICAL SCIENCE COMMUNICATION

CHAPTER 1

THE PUBLIC UNDERSTANDING OF SCIENCE

Jane Gregory and Steve Miller

1.1 Introduction

There have been many episodes in the history of science when scientists have taken an interest in communicating their ideas to the public. During the eighteenth century, for example, the period now known as the Enlightenment produced a wealth of scientific books, many of which sold very widely. Science even became fashionable, and public lectures attracted audiences of everyone from society ladies out on the town to working men looking to improve their lot. At one time, the height of chic was to hire a mathematician to come to your dinner party and perform some differential calculus.

These days we would be hard pressed to find anyone who considered calculus an after-dinner entertainment. At many dinner parties, conversing about last night's theatre or the latest photography exhibition would be taken as a sign of cultured intelligence, while discussing science might be seen as the mark of a bore. This is a problem particularly in the Anglo-Saxon world, with its intellectual divide of arts and sciences into the 'two cultures'—a phrase coined by the author and scientific civil servant C P Snow in 1956. Now, as we move from the twentieth to the twenty-first century, science is both increasingly inaccessible—the volume of scientific knowledge is so vast that no-one could claim to have a grasp of more than a few fields of knowledge—and also trivially familiar: high-tech is in the home. Science may not crop up often at dinner parties, but it crops up all the time everywhere else.

But the ubiquity of science does not necessarily provide an adequate explanation of just why the public understanding of science should be of particular concern to scientists now. Some commentators believe that this concern is a reaction to the secrecy and accelerated pace of change brought about by World War 2; others that it is a counter-attack on the cults and parasciences of the New Age. Some claim that it is a response to an economic demand for technologically strong nations; others that in the face of increasing demands on government funds, science must fight its corner to survive. Whatever the reason, the start of what has been called 'the public understanding of science movement' in Britain can be dated to 1985, when a committee of the Royal Society published its thoughts on the matter.[1] The report drew a number of conclusions, among which were:

> Science and technology play a major role in most aspects of our daily lives ... our national prosperity depends on them...
>
> British industry would be more competitive if those who held positions of responsibility had a better understanding of what science and technology can achieve...
>
> There is scope for more science in the media... [and] a strong case for more science in general programmes, and for improving the contact between scientists and journalists...
>
> Scientists must learn to communicate with the public...

The report has had two important consequences. The first was the setting up of COPUS, the Committee on the Public Understanding of Science. This committee, the members of which are drawn from a wide range of professions and disciplines, provides a focus—and some funding—for the effort of scientists, journalists, and others to communicate with the public.[2] The second consequence was that the Economic and Social Research Council funded a nationwide research programme, so that researchers from a variety of fields in the social sciences could bring their expertise to bear on the issues which are now known under the broad title of the public understanding of science and technology.

The Royal Society had looked in its report at the question of public understanding of science and technology from the point of view of how it affects science, and seemed to be guided by the belief that the more people knew about science and technology, the more they would come to value and respect them. The report, its critics said, asked simply for *more* public understanding of science and technology, without giving a thought to what sort of information people might actually need, or how they might interpret or use scientific information once they'd got it. The research that followed therefore looked at the relationship between science and the public from the point of view of both science and the public, and the field is now populated with a broad range of opinions and approaches. This chapter looks at research in the public understanding of science and technology, and at what it might contribute to the work of science communicators.

1.2 Why do the Public Need to Understand Science and Technology?

It can be difficult to argue against the Enlightenment notion that more knowledge—knowledge about anything—is A Good Thing. However, many scholars have asked in what way, if at all, increased understanding of science in particular might benefit the public. After all, science is only one of many ways that have been developed in order to help us understand ourselves and our place in the world.

One response to this question has been the idea of scientific literacy. This term is derived from the idea of basic literacy—a minimum level of reading and writing skills that people need in order to cope effectively with everyday life. Scientific literacy, by analogy, is the basic level of understanding of science and technology that citizens of a scientific and technological society need to survive in and benefit from their social, cultural and physical environment.[3]

There are many arguments in favour of improved scientific literacy.[4] For example, science is part of our culture and heritage, and everyone therefore has a right to scientific knowledge. Understanding nature can be a source of great pleasure—this is, after all, one of the reasons why people become scientists. In a democracy, people make decisions about scientific and technological policy matters every time they vote. Can we trust ourselves to make such decisions if we don't know much about science and technology? Many of the choices people make every day on a personal level require some scientific knowledge: what to eat, how to travel, how to heat their homes, or how to safeguard their health. Without scientific and technological information it can be difficult to make even these comparatively simple decisions.

The public—as taxpayers and consumers—pay for the products of science and technology: according to the science writer Isaac Asimov, 'without an informed public, scientists will not only no longer be supported financially, they will be actively persecuted.' Wider scientific knowledge might also help reduce the enormous cost of treating avoidable diseases, cleaning up after unwitting polluters and providing crash public education programmes in times of crisis. Any nation that relies for its prosperity and security on science and technology-based industry and services needs both a plentiful supply of scientists, and a public which supports the scientific enterprise.

1.3 What do the Public Know about Science and Technology?

The first survey of public knowledge about science and technology was carried out in the USA in 1957, because the National Association of Science Writers wanted an accurate picture of its potential readership. Such surveys are now carried out regularly in the USA by the National Science Foundation,[5] and in 1988 the Economic and Social Research Council funded the first British survey of the public understanding of science.[6] The survey was designed to be comparable to the US surveys, and it included a series of questions which has become known as the Oxford Knowledge Quiz, and which has since been used in surveys in several other countries.[7]

The results of the UK survey were published in *Nature* in 1989. In some areas, the public knows its science. Almost everyone knows that hot air rises, and 86 per cent know that the centre of the Earth is very hot; 94 per cent know that sunlight can cause cancer, 72 per cent that the continents are moving about very slowly on the surface of the Earth, and 60 per cent that the oxygen we breathe comes from plants. Other results weren't so good: 55 per cent think that antibiotics kill viruses as well as bacteria, and 70 per cent think that natural vitamins are better for them than vitamins made in a laboratory. There are always influences from outside science which affect what people believe: for example, while 80 per cent of British people believe that humans developed from earlier species of animals, only 45 per cent of Americans agree with them.[6]

While surveys like this have produced a great deal of interesting information, they have also attracted a lot of criticism. The surveys, says their critics, reinforce a 'deficit model' of public understanding of science, which says that the public is deficient—that it is lacking in scientific knowledge which it ought to have. Since these surveys largely test people's knowledge of scientific 'facts', the deficit model presents a picture of a scientific community which has all the facts, and a public which doesn't have enough. This raises a number of questions. Who decides which facts we are supposed to know? How many of them do we need? What use are these facts to us anyway?

Although science may seem to undergraduates to consist of a mountain of facts, every working scientist knows that there is more to it than that. What, for example, do scientists do all day, and how does this affect the character of the knowledge they generate? The 'scientific method' is difficult to define. Even within physics the difference between the workings of the theoreticians and the experimentalists—between cosmologists and materials scientists, for example—is vast. Although it may not be possible to describe *the* scientific method to the public, it is certainly possible to describe what particular scientists did in a particular piece of work, and this can offer some insight into the nature and value of scientific knowledge.

However, some sociologists of science—people who study the way in which the scientific community functions as a producer of knowledge—have suggested that knowledge attains the status of 'science' not because of any quality intrinsic to the knowledge itself, and not only because of the way it is produced, but also because of the way the scientific community subsequently tests and endorses or rejects it. The role of the community in the drawing of scientific conclusions—replication, peer review—gives scientific knowledge its authority.[8] Thus the structure and workings of the scientific community play a large part in the task of separating that information that deserves to be called 'scientific' from information that doesn't. If the public are to understand scientific information, and if they are to be able to distinguish the genuinely scientific from the only apparently scientific, then they need to understand not only the methods of science, but also the workings of the scientific community.

Scientific literacy, then, has three aspects: it consists of an understanding and knowledge of the facts of science and technology, of the way in which knowledge is produced, and of the way in which the scientific community decides what is and isn't science.[9]

1.4 Who are the Public? Or, What is an Expert?

Most physicists are not embarrassed to admit that they know nothing about biology. When it comes to biology, physicists are laypeople—they know no more about biology than any other member of the public. The same can be said about different fields within physics: a nuclear physicist may happily know little about electronics.

Within science, then, and even within fields within science, the dividing line between 'expert' and 'layperson' is flexible and dynamic. The same goes for the rest of society. Looking at people in general, we can say that everyone is an expert in a few fields and a layperson in an infinite number of others. Most people have some expertise in fields which are close to them: the geography of where they live, for example, or the habits of their pet or the minutiae of their job. People who live somewhere else, have a different pet and do a different job will have different expertise. These two sets of knowledge are equivalent: each is valuable in its own context, and of little value outside of it.

The same can be said of scientific knowledge. There is much evidence to suggest that people acquire, or are prepared to accept, only that scientific information which they need for their own particular circumstances, and no more.[10] Even people working in highly complex (and potentially dangerous) technological environments often learn only what they need in order to fulfil their particular responsibilities in one small part of the process, and they trust their colleagues' expertise in the rest of the process.[11] Understanding often remains specific to the circumstances which produced it, and is not transferred to other situations. For example, people with long-term illnesses become expert in their particular disease, though very little of what they know may be applicable to other diseases or other people.[12] Gamblers can be adept at calculating the odds on a horse race, but they may be unable to transfer their calculating skills to an abstract problem in statistics.[13] In the 1988 survey people were asked the rather abstract question: what does it mean to study something scientifically? Only 3 per cent mentioned the construction of theories and only 10 per cent mentioned experimenting. However, when presented with a specific, concrete problem—in this survey people were asked to choose from a list a method for working out why a drug wasn't working well—56 per cent chose a controlled experiment.[14]

This knowledge, which is sometimes called 'lay expertise', tends to be specific or concrete rather than general or abstract—the opposite of how scientists see their knowledge of science. But there are times when specific, concrete knowledge is what's required, and the scientists' general laws may not be much help. After the Chernobyl disaster farming restrictions imposed in Cumbria were based on an understanding of what scientists thought was the universal behaviour of caesium in soil, but the data used were for clay soil whereas the Cumbrian soil is peaty.[15] Pellets issued to treat irradiated reindeer in Lapland were believed to be universally appropriate for grazing animals, but they had been designed for cows and were too big for the reindeer to swallow.[16] In both cases local people (the lay experts) could have helped scientists plan appropriate measures, had they been asked.

The result from the 1988 UK survey that attracted most press attention was that 30 per cent of people questioned thought that the Sun moved round the Earth. Scientists have since pointed out that the question is a poor one, since when it asks whether the Earth goes round the Sun or the Sun round the Earth, it does not specify the reference frame. But a more interesting criticism has arisen from researchers who have asked: why is it so important that people should know which goes round which?[10]

If we look again at the arguments for greater scientific literacy, the cultural one seems most relevant in this case. The displacement of the Earth from its position as the unique fixed point at the centre of the solar system was a huge intellectual achievement which had profound repercussions for the way in which thinkers have considered the role of human beings in the physical universe. We can admire this achievement, and gain from it some insight into our place in nature. But does knowing that the Earth goes round the Sun influence the way we vote? Does it help us in our daily lives? After all, even people who plan satellite trajectories assume that the Sun goes round the Earth. In our daily lives, the Sun does go round the Earth.

Some recent research has concentrated not on what people know, but on what they don't know.[17] Scholars working in this area have argued that people develop areas of ignorance as actively as they develop areas of expertise. Like the workers in high technology industries who choose not to know because they trust colleagues with particular areas of responsibility, we all make choices about what we will and will not know. Just as some people choose not to know about history or knitting, other people choose not to know about science.

It may also be the case that people know about science but choose not to use the knowledge. For example, it may be more important for somebody to make decisions on moral grounds which make technical details irrelevant; someone else may feel that they can only achieve peace of mind if they follow their 'gut reactions' or 'instinct'. People may reject scientific ideas about themselves and their world in favour of ideas from other areas of thought which function for them just as effectively. This behaviour has worried some scientists, who feel that the scientific enterprise is under attack from what they call 'anti-science'—ideas which seem to run counter to what science has spent centuries establishing about nature. In the early 1990s, concerned scientists in the USA began to discuss this issue in books and at conferences. Worried about science's apparent dwindling authority, the scientists constituted what became known by its critics as the 'anti-anti-science movement', and debated the role of science and its place in the wider culture with sociologists, historians and other scholars. The impact and outcome of this debate remain to be seen.[18]

1.5 What do People Think about Science and Technology?

Despite some scientists' worries, and contrary to the pessimistic expectations of many, current public attitudes to science and technology are encouragingly positive—which can only be good news for science communicators.

Ninety per cent of people think that governments should fund blue-sky research, and 86 per cent think that science and technology are making our lives healthier, easier and more comfortable. However, 60 per cent think that science and technology make our way of life change too quickly, a slight majority is against animal experiments, and a slight majority disagrees that computers and automation would create jobs.[19]

Assessing levels of public interest in current events, surveys repeatedly find that the public say they are very interested in science and technology. One survey found that while 77 per cent of people asked were interested in local issues, 57 per cent in defence and 59 per cent in business news, 92 per cent were interested in medicine and 88 per cent in science.[20] In another survey, people were shown a list of thirteen newspaper headlines, and asked which they definitely would or wouldn't read. Out of topics including sport, politics and showbusiness, science held its own surprisingly well.[6]

Clearly people think that science and technology are important and relevant to their lives. However, when asked whether they were 'informed', the percentage of people who claimed to be informed about science was much lower than the percentage who claimed to be interested in science, while for sport and politics the figures for informedness and interest were about the same.[6] Many people believe that increased knowledge would be a good thing: when asked to respond to the statement 'It is not important for me to know about science in my daily life', 63 per cent disagreed.[19] However, the belief that more knowledge of science leads to more positive attitudes towards science turns out to be misplaced: analysis of the 1988 survey data shows that people who know more about science have more sharply defined views about it—but these views can be negative as well as positive.[21]

Many of scientists' feelings about the public's perception of them as people are unjustified. Scientists are generally held in rather high regard. Surveys show that the public thinks that scientists are responsible, concerned and not particularly different from anyone else. They are no more or less sociable, arrogant or eccentric than other people, but they are more secretive than average.[20] The feeling that scientists are secretive is an interesting reflection on the access laypeople have to scientists and scientific knowledge. While some scientists may need to be secretive, many will be unaware of just how inaccessible they are to the public.

1.6 How can Communication Help?

So the public's attitude to science and technology is positive, but their knowledge of science and technology is limited. There is nothing particularly extraordinary about the low levels of knowledge of science: most people don't know much about politics, history, literature or foreign languages either. As one researcher has pointed out, however surprised we might be about these measures of knowledge, we should assess them in contexts such as that provided by his home state, Kentucky, where 20 per cent of the population are illiterate—they cannot read and write.[22] However, the fact that the public knows no less about science and technology than it does about a number of other important fields is no reason to be content with the situation, particularly since a significant proportion of the public is keen to know more.

But how can research into the public understanding of science and technology improve the work of science communicators, and how can science communicators improve the public understanding of science and technology?

Since the time of the Renaissance the sheer volume of scientific knowledge has increased enormously, as has the range of subjects which are now considered suitable for scientific investigation. The trend, particularly in this century, has been for scientific knowledge to become more quantitative, with the application of mathematical methods to every branch of science and technology.

While people in developed countries today have increasing contact with the fruits of scientific knowledge, the tendency towards quantification and specialisation has erected barriers which have to be overcome—scientific language, for example, can make scientific ideas seem more complicated than they need be. But all scientific knowledge is knowledge about the universe in which we all live. The public's eagerness to claim their share of their scientific heritage is evidence that there is no *a priori* reason why scientists cannot contribute to improving public scientific literacy: doing so is simply a question of will and technique rather than divine inspiration.

If we look at the three-fold requirement of scientific literacy, we can see several ways in which the science communicator can contribute:

- The first aspect of scientific literacy is an *understanding and knowledge of the facts of science and technology*. Most science communication involves the communication of the facts of science and technology. However, science communicators could bear in mind that people find most useful, and most readily retain, information which is relevant to their lives and which is set in a real, everyday context rather than in the abstract terms of formal science.

- The second aspect of scientific literacy is an *understanding and knowledge of the way in which scientific knowledge is produced.* Science communicators could bear in mind that the people who produced the knowledge and what they actually did are important components of any scientific story (and also offer evidence of the human side of science and technology).
- The third aspect of scientific literacy is an *understanding and knowledge of the way in which the scientific community decides what is and isn't science.* Science communicators could bear in mind the implication of an endorsement from the scientific community for the status of scientific knowledge, and could mention, for example, publication in a peer-reviewed journal or the opinions of other scientists.

References

1 Royal Society 1985 *The Public Understanding of Science* (Report of The Royal Society, London)

2 For information about COPUS contact the COPUS office at The Royal Society, 6 Carlton House Terrace, London SW1Y 5AG

3 Durant J 1992 'What is scientific literacy?' *Science and Culture in Europe*, edited by J Durant and J Gregory (London: Science Museum) 129–137

4 Thomas G and Durant J 1987, 'Why should we promote the public understanding of science?' *Scientific Literacy Papers*, edited by M Shortland (Oxford: Department for External Studies, University of Oxford) 1–14

5 See Miller J D 1992 'Toward a scientific understanding of the public understanding of science and technology' *Public Understanding of Science* **1** 23–26

6 Durant J R, Evans G A and Thomas G P 1989 'The public understanding of science' *Nature* **340** 11–14

7 See for the US and UK: Evans G and Durant J 1989 'Understanding science in Britain and the USA' *British Social Attitudes: Special International Report*, edited by R Jowell, S Witherspoon and L Brook (Aldershot: Gower) 105–120; for other countries and international comparative data contact the Public Understanding of Science Research Group, Science Museum Library, London SW7 5NH.

8 See Barnes B 1985 *About Science* (Oxford: Basil Blackwell) and Collins H M and Pinch T 1993 *The Golem: What Everyone Should Know about Science* (Cambridge: Cambridge University Press)

9 Shapin S 1992 'Why the public ought to understand science-in-the-making' *Public Understanding of Science* **1** 27–30

10 Lévy-Leblond J-M 1992 'About misunderstandings about misunderstandings' *Public Understanding of Science* **1** 17–21

11 See the work of Brian Wynne, University of Lancaster, on the Sellafield apprentices.

12 Lambert H and Rose H 1990 'Disembodied knowledge? Making sense of medical science' Paper presented at the conference *Policies and Publics for Science and Technology*, Science Museum, London, 7–11 April 1990

13 Paulos J A 1988 *Innumeracy* (New York: Hill & Wang)

14 Durant J R, Evans G A and Thomas G P 1989 'The public understanding of science' *Nature* **340** 11–14

15 Wynne B 1992 'Misunderstood misunderstandings: social identities and public uptake of science' *Public Understanding of Science* **1** 271–294

16 Paine R 1992 '"Chernobyl" reaches Norway: the accident, science, and the threat to cultural knowledge' *Public Understanding of Science* **1** 261–270

17 See the special issue dated December 1993 of the journal *Knowledge*.

18 See, for example, Holton G 1993 *Science and Anti-Science* (London: Harvard University Press) and Gross P R and Levitt N 1994 *Higher Superstition: the Academic Left and its Quarrels with Science* (Baltimore, MD: Johns Hopkins University Press)

19 Evans G and Durant J 1989 'Understanding science in Britain and the USA' *British Social Attitudes: Special International Report*, edited by R Jowell, S Witherspoon and L Brook (Aldershot: Gower) 105–120

20 Shortland M 1987 'Networks of attitude and belief: science and the adult student' *Scientific Literacy Papers*, edited by M Shortland (Oxford: Department for External Studies, University of Oxford)

21 Evans G and Durant J 1995 'The relationship between knowledge and attitudes in the public understanding of science in Britain' *Public Understanding of Science* **4** 57–74

22 Fullwinder R K 1987 'Technological literacy and citizenship' *Scientific Literacy Papers*, edited by M Shortland (Oxford: Department for External Studies, University of Oxford)

CHAPTER 2

GATHERING INFORMATION

Jane Gregory and Steve Miller

2.1 Introduction

Even if you are dealing with your own scientific specialism, you will find when you come to write a paper or give a talk that there are gaps in your own knowledge. At this point you will have to draw on the collective knowledge of scientists in your chosen area. This knowledge is to be found in books and museums, and it is updated continually in scientific journals and in popular science articles and broadcasts. Specialised institutions and organisations may also be able to help, and don't forget that scientists themselves can often be a very useful and lively source of information. The range of possible sources is wide, and in this chapter we look particularly at those which are useful in researching talks, articles and books intended for popular—or at least non-specialist—consumption.

Thorough and accurate research adds variety, provides a sense of the community of scientific work and gives you confidence and authority; it helps you ask the right questions when you interview your sources, ensures that you don't miss out on a vital piece of information, and helps you to answer any questions that your work might prompt.

In the end, what you select from and make of the wealth of scientific information at your disposal is up to you. But try to remember: all scientific knowledge is the result of the painstaking work of generations past and present. It may have been hard won; people may have suffered in their efforts to obtain the information now so readily to hand. So treat it with respect: select information appropriate to your purpose and don't abuse or distort it to fit with your particular preconceptions or prejudices.

2.2 Libraries

2.2.1 Books and Journals

Undergraduates spend most of their time acquiring the knowledge they need to pass exams, and—by and large—this knowledge is provided in lectures. 'If it's not in the lecture it won't be in the exam', students often argue; they may therefore see little reason to use their library. However, if you are going to write a thesis or a scientific paper, not only do you need a good background understanding of your subject, but you also need to be up to date with the latest research. A visit to the library at least once a month is essential. If your aim is to write popular science articles you need not be quite so religious, but you will have to check your article for basic scientific accuracy.

Academic libraries offer textbooks, reference books, and journals of current research and their back numbers.

Textbooks

Textbooks provide a broad, structured, accessible summary of a single topic. They are written to accompany a course syllabus, so their content is defined by the level and syllabus of the course, rather than by the subject. So do not think that they represent the subject in its entirety—you will have to look elsewhere for that.

Reference Books

These can be extremely specialised monographs, classic works by leading scientists, encyclopedias of data, or working handbooks. Undergraduates sometimes think that science is very cut and dried—that there is only one answer—but the wide range of reference books available is evidence that this is not the case. Explanations of the same phenomenon by different authors may differ considerably, so it is worth reading around to find the one that best suits your purposes. Histories and biographies may also yield interesting information about people, places, and the development of ideas, and will offer an insight into the contexts—social, economic, political, technical—in which science is done.

Journals

Academic science libraries may carry hundreds of journal titles. Only some of these will be relevant to your work, and you should ask advice on which you should read regularly. Most specialist journals publish papers which describe new research. They have a very formal structure and style, and assume their readers have a high level of specialist knowledge. Some journals, which are called review journals, publish articles which bring together recent work in a particular subject. Review articles are generally more readable than research papers, and can be a useful starting point for specialist and popular articles. In addition, many journals have a 'letters' section (this may even be a separate journal, as is the case with *Physical Review Letters*). Letters to journals are not the same as letters to newspapers: in a journal the letters are short articles which are published quickly because they contain information which is particularly useful, interesting or important. Authors of letters often write a full paper later.

Within a particular journal, only a small number of papers will be of direct interest to you. The journal itself, or the contents list, may be subdivided into narrower sections. With a little practice you will be able to scan the list of papers quickly for relevant material. Then it is worth reading the abstract to see if the paper really does contain information which is useful for you. Letters, research papers and review articles end with a list of references to previous work. These references trace the intellectual heritage of the work, and can lead you to its precursors. They are also extremely useful as a general guide to journals you should check regularly.

Journals are usually the first place where new work is presented to the scientific community. Publication of a paper in a journal means that the work has been through the peer review system. Peer review is the process by which the scientific community checks and either rejects or endorses the new work. Any paper submitted to the journal will be sent in confidence to several 'peers'—anonymous, independent referees who are experts in the appropriate field—and they will check to see whether it is accurate, interesting and original. The paper will only be published if it meets these criteria. So although the results may have been announced before, at a conference for example, their appearance in a peer-reviewed journal means that you can use them in material for the general public with a reasonable degree of confidence that what you are reporting has the support of the relevant section of the scientific community.

Two journals which are particularly useful are *Nature* and the American journal *Science*, since they frequently publish newsworthy papers, often with commentaries by eminent scientists. However, if you want to concentrate on popularising just one field of science, you should also monitor the relevant specialist journals.

2.2.2 How to Find Information in a Library

Libraries can be baffling. Faced with floor upon floor of wall-to-wall knowledge, how do you find the bit you want?

Libraries have an indexing system which is called a catalogue. The catalogue is sometimes divided into time periods (for example, books published before and after certain dates), subject areas, or types of publication. The catalogue may be split between different systems: some may be on computer, some on microfiche, and some on cards. Publications may be listed under title, or author, or both. Because a library's catalogue can be confusing for the newcomer, you can save yourself a lot of time by asking the librarian. They may give you a written guide to the library, or offer you a guided tour. Librarians know what information is available, and it is their job to help you find it. They are an undervalued and underrated species, and you should take full advantage of their knowledge of the institution where they work.

Public Libraries

Local public libraries are often repositories of local history. If you were to research an article on chemical pollution in the Wirral, a stop at Runcorn or Widnes Public Library would be well worth the effort. They will probably have back copies of the local paper, which would have covered any leaks or scares from the local chemical industry in far greater depth than the national press.

There will be books with photographs of what the area looked like prior to the arrival of the local power station.

Local libraries will also have information following the careers of people who have reached national or international fame. 'Brixton hockey star wins Cambridge scholarship' may fill a page in the *South London Press*, without making a ripple in *The Times*. But the former hockey star may go on to win the Nobel Prize in later life, and Brixton library's back issues of the *South London Press* will be a good place to start gathering background information for a profile of the new Nobel laureate.

If the information you want is not actually in the library itself, the noticeboard may have details of the next meeting of the Isle of Wight Astronomical Association or the Royal Society of Scotland's Science and Engineering section, together with the phone number of the local secretary. And—once more—why not ask the librarian if they can help you find that council report on the threat to local wildlife caused by the Channel Tunnel Rail Link?

Picture and Photograph Libraries

In chapter 3 we outline the importance of illustration for making science communication both easier and more attractive. But that often means finding just the right picture. Maybe you've already seen just what you want somewhere, but the magazine got thrown away just before you needed it. Picture and photograph libraries may be able to help, and you can find them by looking at the credits in illustrated books and magazines. Although you probably won't be able to wander in and browse like you can at book libraries, many photo libraries still welcome visitors. Most of them are run commercially, and the fee they charge may well depend on what you want to do with the picture. If you need something for a nationwide advertising campaign, you will usually pay more than you would for a slide for a talk to the local primary school.

If you can assure a photo library of your business or academic credentials, they may send you a selection of prints or slides to look at. Some libraries charge a fee for digging out a relevant selection, but most just charge you for the pictures you use. Many museums, galleries and planetaria have slide collections that may contain just what you've been looking for.

Cuttings Libraries

Cuttings libraries hold articles from back issues of newspapers, sorted according to subject. These libraries are also usually run commercially, but can be very useful if you want to make sure you have read all that got into the press on a particular event. Fees can run high: you can be charged per article, for photocopying, for the librarian's time, or even for the time you spend in the library rooting around for yourself. If you know that the article you want was in a particular issue of a particular newspaper, it may be cheaper to buy a copy of the whole paper from the publisher's back issues department.

2.3 Live Sources of Information

2.3.1 Living Science

Science is a human activity carried out by living human beings. Yet the scientists who research and teach science are often overlooked as sources of information. One of the difficulties of gathering information on many topics of scientific interest is the sheer volume of literature on the subject, but half an hour spent with someone active in that area of science may save hours—if not days—of your time.

Before you intrude on the time of busy scientists, be well prepared. The person you are going to see may also need some time to get ready. So:

- As far as possible, arrange to meet at a time which is mutually convenient, and at a place which is suitable for one-on-one conversations. It is possible to carry out a meaningful interview on the bus in the middle of the rush hour, but not advisable.

- Try to find out something about your topic on your own—at least enough to draw up a list of sensible questions. Unless you really are writing about the theory of everything, make sure your questions are reasonably focused on your immediate topic of interest.
- Make sure you have allowed enough time to ask all you want, plus a little extra in case something takes longer to understand than you expected, or a really interesting, but unforeseen, point comes up that is worth pursuing further.
- Take with you to the interview your list of questions, a notebook and at least two pens or pencils. A tape recorder (with tape and live batteries) is often very useful, especially if your scientist is a fast talker; and it will make it easier for you to quote the scientist directly.

It is particularly useful to talk to an expert if you have to write or talk about a subject which is relatively foreign to you. A scientist who is active in your chosen topic can act as a 'talking book', and will be able to

- give you a general introduction to the subject;
- point out areas of controversy, highlight important gaps in current knowledge, and tell you which are the most interesting recent results;
- give you a guide to the literature, or at least point out where you might start to find the information you require;
- comment on the work, discoveries or claims of other scientists working in the field.

When you use your scientist as a talking book, you are not interested in the exact words or phrases they may use: they are giving you general information.

Make sure you have the right person: the wrong one may be flattered into answering your questions on a subject on which he or she is not qualified to speak. You may be able to find a suitable scientist by yourself, but it is worth knowing that the CIBA Foundation runs a Media Resource Service for journalists (and other interested parties) which puts them in touch with an international panel of scientists who can be used in this way. The Media Resource Service's telephone number is 0171 631 1634.

2.3.2 Interviewing

It may be that you are interested in a particular scientist as well as in his or her area of expertise. They may just have made an important discovery or been given a prestigious post or award. In this case you may want to carry out a formal interview, and note the actual words or phrases they use for quoting directly—or at least only partially paraphrased. The whole of what you write may simply be an interview with the 'person of the moment'. At such times, your scientist is likely to be even busier than usual.

You will need to be thoroughly prepared. As well as having enough knowledge of your subject to formulate some interesting questions, you will need to know something about the person you are interviewing. Various sources may be used for this. Your subject may be important enough to warrant a mention in *Who's Who*. They may belong to a professional body which can help you; or you could ask at their place of work for some biographical details.

When interviewing someone you should

- Draw up a list of all the points you want your scientist to discuss. Then try to cut this down to around a half-dozen questions which should give your interviewee the scope to cover the relevant ground.
- Avoid questions which could be answered with a straight 'yes' or 'no'—you need to get your interviewee talking. Avoid lengthy questions which are designed to show off how much you know.
- If the result of your interview is to be an article for public consumption, try to stick to the main points of the science concerned. Minor details may be fascinating to you, but they will leave everyone else cold.
- At the outset, try to establish rapport with your subject. This depends very much on your behaviour: be on time, reasonably smart, friendly and awake.

- Unless you are going to be deliberately adversarial, don't set out to antagonise your interviewee. Keep the most controversial question until last or you may end up with a very short interview.
- If, during the interview, the conversation is heading for a dead end, tactfully interrupt and direct it toward where you want to go. If it has taken an unexpected but interesting turn—the whole point of doing an interview is that the scientist knows things that you do not—let it run, at least for a while.
- Make sure you understand what your interviewee is saying. If you don't, ask them to go over things again.

Writing up the material from an interview requires both brutality and a lightness of touch. You may wish to write most of the material in reported speech form, and incorporate background material. Another way to write up an interview is as a straight question-and-answer piece.

Some words of warning:

- Do not put words into your interviewee's mouth or distort what they have said.
- Do not take individual words or phrases out of context.
- Do not ignore the qualifications that a scientist may add to their statements if doing so makes them look too sure or narrow-minded.
- Do not alter your questions after the interview to make your interviewee look ridiculous or evasive.

When writing up an interview, you may

- change the odd word to avoid repetition and leave out words that people use while they gather their thoughts (well, um, er);
- re-order material to help the flow of the article;
- remove obscuring, qualifying material if it hinders understanding—you may be able to use it later;
- alter your questions to help the flow of the article.

As interviewer, you are acting as ambassador for your institution or for the publication in which you hope to place your article, so you should always be pleasant to everyone with whom you have to deal. If you are rude to the Professor's secretary they are unlikely to squeeze you into that slot normally reserved for emergency meetings. If you are a professional scientist, you may find fellow scientists more willing to talk to you than they would be to full-time journalists—but no-one owes you anything.

2.4 Other Sources of Information

2.4.1 Museums

Behind the public face of most of our national and local museums you will find top quality research programmes on the history and contemporary practice of science and technology. This research will be reflected in the exhibits, especially if the museum concerned is putting on a special display of the latest information on a particular subject. If all you need is not immediately available, the information desk should be able to give you the name of the curator responsible for your area of interest and tell you when they are available to talk to you. If you are not able to make a personal visit, a telephone call, followed by a courteous letter giving precise details of what you want, will often get you the information you seek.

2.4.2 Professional Organisations

House magazines have a reputation for being as dull as ditchwater. But many professional bodies and science-based companies now produce high quality publications which give detailed information about their work. If you want some information about North Sea oil rig construction, British Petroleum may have an article in one of their company magazines which is just what you want. For balance, it may be that Greenpeace have also had something to say on safety features or procedures for dealing with spills.

Professional organisations also have a good idea of who's who in their area. For example, a call to the Royal Astronomical

Society will get you a list of half a dozen of the country's leading astronomers who also take a keen interest in popularising the subject. And if the Institute of Physics can't tell you who to contact for the latest thinking on the Large Hadron Collider, no-one can. For the latest information on the long-term effects of the Chernobyl disaster, you should not ignore the Farmers' Union—not a normal port of call for a story about radiation, but in this case a key one.

2.4.3 Opinion Research Organisations

Just what does the public think about the introduction of the latest genetically engineered vegetables? Can you really taste the difference between butter and the latest synthetic offering from the chemical industry? Opinion pollsters have the answers. And they often have to ask the same question year in, year out, making their information valuable in establishing trends. You may have to pay for the information, but if a million dollar contract hangs on your being able to demonstrate that tests show dogs really would choose reconstituted old boots over prime steak, market research companies may have been there and done that ahead of you.

2.4.4 Computer-based Sources

Nowadays no academic institution is complete without being connected to the world-wide web (WWW or web, for short), a global network that links computers via cable and satellite. These linked computers are based not only in colleges and universities, however. Many government agencies, political parties, businesses, newspapers and homes also now have networked computers of one sort or another: in the United States, some 50 million people can access the web from their own homes. To gain access to this network—the internet—you either have to have an account on an already networked computer, or have your own personal computer (p.c.) or workstation registered as a recognised 'address' to which electronic mail can be sent. Either way, you have to run some computer program which acts as a mailer, and your p.c. or workstation has to have a piece of hardware known as a 'network card' and be connected by phone or cable to the outside world.

The resulting network is what makes it possible for computer users to send electronic mail, or e-mail, to their colleagues and friends at any time of the day, leaving urgent messages, exchanging information and, often, just chatting. Many a fruitful collaboration has been started by e-mail, long before scientists have had the chance of a physical encounter. E-mail can be a way to get information 'hot off the press' and of brainstorming to good effect. But be considerate. To save yourself a trip to the library, where the information you want may be freely available, you could be wasting other people's time: many scientists are hopelessly overloaded with requests for information, and, in self-defence, they may consign a good proportion of their daily messages to the electronic dustbin without even glancing at them. You should treat e-mail as if you were sending a letter, in terms of politeness, although the medium does allow for greater informality than real paper.

An extension of e-mail are mailing lists and electronic bulletin boards. Mailing list subscribers passively receive centrally generated messages: people interested in astronomy, for example, may subscribe to NASA's or the Royal Astronomical Society's electronic press release system. Scientists interested in various subjects have set up lists of colleagues and subscribers with similar interests who can actively post messages to a bulletin board. These messages are then sent out as e-mail to everyone on the list, often generating discussion among the board members. Some of these boards are moderated; an 'editor' looks at incoming messages and decides whether or not to post the offering. For others, no such editing takes place: what arrives on the board goes straight out to the subscribers. That means that the quality of information varies. While some boards are well-focused and disciplined, others are liable to generate vast quantities of low-quality messages of only marginal interest to the majority of subscribers. Know what to join and when to leave.

For those interested in the public understanding of science, a bulletin board of interest is PCST-L, run by Cornell University. Its workings are typical of such boards: to subscribe, you send a one-line e-mail message 'subscribe PCST-L firstname lastname' to listproc@cornell.edu, where you put your first and last names in the appropriate places. You will then be able to receive messages sent to the board and make your own contributions. The 'protocol for using PCST-L' is a useful guide for the responsible world-wide web user and is reproduced in figure 2.1.

In addition to bulletin boards, you may access databases and 'expert networks' for information via the web. Databases may consist of useful catalogues of research publications, such as the BIDS bibliographic database and citation index run by the University of Bath, or they may be databanks, such as the HITRAN list of infrared spectral lines managed by the United States Airforce. Your central computer may have access to only a limited number of such databases; if the one you really need is

GUIDELINES FOR PCST-L MESSAGES

These guidelines are intended to make PCST-L a productive region in cyberspace. As experienced users know, the signal-to-noise ratio of electronic discussions can be poor; by following these guidelines, we can filter out some of the worst noise. So, keep in mind:

- When replying to a post, please keep quoted material to a minimum; often a couple of lines are sufficient to get the idea across. Please do not quote the entire post you are responding to; edit it.
- One-liners and short quips generate lots of noise. Please take care to contribute something of substance; more than 'I agree' or 'I do that too'. It's nice to know, but in the end generates lots of messages without actually discussing much.
- Include your e-mail address and name at the end of your post. This is often called a signature. Please keep your signatures short—three or four lines at most.
- Advertising is discouraged, although mention of commercial products may at times be appropriate in some discussions.
- The following are prohibited: off-topic posts; posts which are excessively inflammatory, derogatory, or encourage violence; administrative queries and comments; posts seen as causing disruption.

Figure 2.1: *Guidelines for using an internet bulletin board.*

not there, ask your computer manager or head librarian if it could be installed or subscribed to. Real money may be involved in these subscriptions, so make sure you know what is in the database and ask yourself if you really do need the information it contains. For specific information you may wish to contact an expert network. In the United States, there is 'profnet' (profnet@vyne.com). Networks for the press include the Media Resource Service run by the Novartis Foundation (mrs@novartisfound.org.uk).

Far more widespread than any of these resources, now, are so-called 'web pages', online information including graphics, video and sound clips, posted by individuals and organisations. These may vary from highly informative services, such as those providing detailed weather maps, to little more than advertisements or statements of personal belief. To access web pages, your computer must be running a web access program such as Netscape. Web pages have specific addresses; to find out about the Galileo Mission to Jupiter, for example, you go to the relevant page run by NASA's Jet Propulsion Laboratory in Pasadena, http://www.jpl.nasa.gov/galileo/. If you do not know the exact web address, you may use a net search engine, like AltaVista (http://www.altavista.digital.com/) which allows you to input key words, refine your search and get the best matches. Many journals are now available over the web, provided that you or your institution has the relevant subscription; the general journal *Science*, for instance, may be reached at http://www.sciencemag.org/.

While it has enormously speeded up the rate at which information can be publicly accessible, the net, and particularly using web pages, has many drawbacks. Except in the case of online academic journals, very little of what you find has been peer reviewed, and you need to be cautious, particularly where claims are large and controversial. There is a democracy on the web: everyone can have their say in this unregulated world of electronic information. And because it is freely accessible to anyone with a networked computer or access to one, the net generates an enormous amount of electronic traffic; downloading information from an American site in the British afternoon, when the East Coast of the USA is awake, can take a very long time. Sometimes a trip to the library to look at the printed journal is quicker.

The information superhighway must be travelled with care. It is possible to spend forever surfing the net, gathering fact after fact, without ever becoming any the wiser. If information can travel across the network, so can computer viruses, so scan and clean your machine regularly. And finally, at some point you will have to stop gawping at what everyone else has done, and get on with some work of your own.

2.5 Copyright and Acknowledgements

If you are going to make use of information gathered from any of these sources you may run into problems with copyright. You must obtain permission to use other people's material. Images of Saturn from NASA or particle tracks from CERN may come free: these large organisations are often happy to receive a credit in the picture caption. Other sources may require payment before you can reproduce their material. Your institutional or local library, or your publisher, should be able to advise you on just what you need to do to ensure you are not breaking the law. Copyright nowadays also covers the use of computer programmes and routines. This is a particular problem if you are trying to make money out of a package that makes heavy use of 'off-the-shelf' software.

You should always check on the conventions for acknowledging sources when you quote or report in print the words or ideas of others. People who have been very helpful should receive appropriate thanks, or they may be less willing to help next time. Make sure also that you do not compromise those who have assisted you. If you have written a *New Scientist* article which makes use of material to be published in *Nature*, make sure your article does not appear before *Nature* hits the street. *Nature* has strict embargoes and will refuse to print an article that receives advance publicity. And always check whether your institution has any regulations regarding what employees can publish.

CHAPTER 3

WRITTEN COMMUNICATION

Jane Gregory and Steve Miller

3.1 Introduction

> 'Most of what I do is essay writing and I do it for different reasons. In common with most essayists, I really do it for myself. If I did it for any other reason it wouldn't be any good.'—*Stephen Jay Gould, palæontologist and popular science writer.*

Certainly a science writer must write to communicate—Gould would be the first to admit that one of his essays does not work if it does not communicate—but fundamentally, he insists, 'you write essays to try and write the perfect one, and since you never get there you have to keep doing it, and that's why it's done'.

This implies that good science writers are at heart artists, working tirelessly for themselves, aiming—perhaps forlornly—to hone their innate talent to perfection. So maybe you have to be born a writer. But if you read a Stephen Jay Gould book, or one of his essays for that matter, you will immediately be impressed by his craft skill. Gould uses rules and techniques which govern the way he writes, and they are what make his writing work. What follows here is based on the belief that everyone, no matter how little innate talent for writing they may feel they possess, can benefit from knowing about and applying some basic rules of writing.

Some people think that ideas expressed in clear, simple English must be simple ideas; and that ideas expressed in complicated language must be difficult ideas, which are somehow more valuable than simple ones. These people are mistaken. In order to communicate we need to make ideas seem clear and simple, however difficult they are—and to do that we must write in clear, simple English.

Your work should bear as much as possible on the interests and experiences of your readers. Since you will not be around to help each reader make sense of what you have written, you must be accurate and precise. These qualities are especially important when you are dealing with technical matters or with matters which, if misunderstood, could create anxiety or harm.

Some common-sense rules apply, whatever you are writing:

- Have a clear and specific idea about what you are doing: you should be able to state the purpose of your writing explicitly in a single sentence.
- Decide who you are writing for. What do your readers already know, and what do you need to tell them? What will your readers expect your work to look like?
- Be clear about whose interests you represent when you write: your own, your company's or research group's, your profession's? What would these people expect your work to look like?
- Know your medium: study good examples of the type of writing you are trying to achieve, and analyse their form, style and content. Ask for guidelines—most journals and some publishing houses have them—and follow the guidelines to the letter.

3.2 A Guide to Clear Writing

All forms of written communication require clear and simple English, uncluttered construction, and good organisation. If you are writing a paper for the specialist scientific literature, your work is likely to be read—and refereed—by people for whom English is a foreign language, so clear straightforward writing is especially important.

Anyone wishing to write clearly can learn a lot from reading a tabloid newspaper once in a while. If you look at the way an article is written for the tabloids, ignoring the content of the story and the paper's own form of jargon, you will see that the story is usually expressed in a concise and immediate fashion. A lead article may be only 300 or 400 words—20 or 30 sentences—but it will contain all the information the writer wishes to convey. Other features of a tabloid article are:

- The sentences are generally short, averaging about a dozen or so words. They have few sub-clauses (sub-clauses are marked off from the rest of the sentence by pairs of commas, brackets or dashes).
- The verbs in a newspaper article tend to be active: 'The man bit the dog' rather than the passive 'The dog was bitten by the man'. Active verbs tend to make the sentence shorter and more immediate.
- Newspapers use short, everyday words. These are always better than long, complicated words. Long words are a particular problem in science, where there is so much jargon.

Include these features in everything you write, and your work will instantly become clearer.

3.2.1 Sentences

Here is a sentence taken from a student's attempt to explain some of the latest research carried out by Gaia scientist James Lovelock and his collaborators:

> The researchers designed a model to compare temperature regulation with the extent of plant life on land and marine algae, finding that if carbon dioxide rises continued unchecked, and the average global temperature increases to 20°C, as some climatic models predict it might by the end of the next century, the land and ocean life forms would amplify the increase and encourage a transition to a permanent warm state, much hotter than today, similar to the change which is believed to have ended the last Ice Age to give us our comparatively mild interglacial climate.

Only a South Pacific pearl diver would not be out of breath after reading that sentence aloud. It is 95 words long and contains everything except the kitchen sink. But it is not just the length of the sentence that is the problem here: simply too many ideas are strung together, and there are few of the breaks that careful punctuation would provide to allow the reader to follow the argument easily. The information is densely packed: what does the phrase 'compare temperature regulation with the extent of plant life on land and marine algae' really mean? But with a better sentence structure and a little less haste, it is possible to rescue all the ideas contained here. One might try:

> The researchers designed a model to mimic the way in which the Earth's temperature is regulated by plants on land and algae in the oceans. Some climatic models predict that if carbon dioxide levels continue to rise unchecked, the average global temperature might increase to 20°C by the end of the next century. In that case, the Lovelock model predicts that land and marine vegetation would amplify the temperature increase, encouraging a transition to a permanent warm state. A similar change is believed to have ended the last Ice Age, giving us our comparatively mild interglacial climate.

The rewritten version is also 95 words long, but it is divided into four sentences and the information comes in much more manageable pieces. Now the piece feels less rushed. Composite nouns like 'temperature regulation' have gone and, because each main idea is contained in its own sentence, one can, for instance, distinguish between 'climatic models' in general and the researchers' model in particular.

Keep your sentences as uncluttered as possible, and check to see that nothing in them is needlessly wordy or redundant. An additional incentive for this is that many publications insist you write no more than a specified number of words to meet their requirements. Newspapers are renowned for cutting articles submitted to them, often to the extreme annoyance of the author. If you wanted to write a Letter to *Nature*, for instance, you would find a rigorous 1500 word limit placed on your publication. While the referees may demand more scientific information, *Nature*'s editors will insist that your paper is concise and to the point. By getting rid of clutter, you can satisfy both.

Clearing out unnecessary verbiage will not only help you keep within the required word length. The result should be that the whole piece will also read better. For example, the four words 'a considerable amount of' could be shortened simply to 'much', saving you three words. 'In the interim' could be replaced by 'meanwhile' and 'in the event that' by 'if'. There are many other such examples which you should look out for.

You should also see if it is possible to remove words or phrases from a sentence without affecting its overall meaning. 'The nature of Hoyle's work is always of a provocative kind' could be replaced by 'Hoyle's work is always provocative'.

There is redundancy in the phrase 'the deposited precipitate' because of the similarity in meaning between 'deposit' and 'precipitate'. You don't need to say 'the given data' because 'data' means 'what are given'.

Don't waste space by attempting to qualify words such as 'unique' or 'absolute'. 'Nearly unique' should not be used; the phrase you want is 'extremely rare'. 'Almost absolute' is still 'relative', and since you cannot get better than 'absolute', 'completely absolute' is a waste of a word.

3.2.2 Words and Jargon

Within the scientific community there appears to be a feeling that if you say something in a complicated, slightly unfamiliar way, your words acquire more weight and are taken more seriously. Depending on who your intended audience is, this unnecessary complication can be very off-putting. One of the reasons why many people find science difficult to understand is that scientists seem to have invented a language of their own, or have given entirely new meanings to everyday words.

Let's start with individual words. In 'Jupiter's diameter is *approximately* 140,000 kilometres' the word 'approximately' could quite happily be replaced by 'about'; in 'the dinosaurs became extinct *approximately* 65 million years ago', 'around' would do the job. Why do we have to '*demonstrate* that an Arrhenius plot fits our data adequately'? Why can't we just 'show' it? 'The *optimal* fit to our line profile is a Gaussian, the Lorentzian having been rejected as *sub-optimal*' means simply that the Gaussian fit is the 'best' and the Lorentzian is 'not so good'.

In day-to-day affairs, people are used to being 'charged' by shops for the goods they buy—and some may have been 'charged' by a bull. Everyday objects like vacuum cleaners and alarm clocks have 'parts', which never go back together again when you have taken them to bits. But 'charged particles' are a bit of a mystery, although a non-scientist may feel that the words have a familiar look to them. 'They should lock so-and-so in a *cell*, and throw away the key', is a familiar cry. But a 'cell' to a chemist is a source of electrical energy; to a biologist it is the basic building block of living things.

Scientists have got into the habit of stringing together nouns to produce composites which carry a lot of information to the specialist in that particular field, but mean little to anyone outside the chosen few. 'Here we see a *typical late type asymptotic giant branch power spectrum*' is not only opaque to the uninitiated, but it's also almost impossible to say. At the very least one needs some commas or hyphens: 'typical late-type, asymptotic giant branch, power spectrum' lets you know which bits go with which. In this phrase, the final noun 'spectrum' is being qualified by adjective, adjective, noun, adjective, adjective/noun, noun, noun, so it is no wonder that it takes some sorting out. This type of jargon is very useful among specialist astronomers (in this case) for packing a huge amount of information into very few words. For the general public, however, you would have to write a feature-length article on the ways stars evolve if you wanted to start to get across the ideas contained in those eight words.

In many walks of life, people define their professional affiliations by which sets of initials, which abbreviations, convey meaning to them. Science is full of abbreviations, which are another form of jargon. Some abbreviations have come into such general use that they need no explanation. The American space agency is now so well-known that it is often written Nasa, rather than being fully capitalised. Who remembers what 'laser' once stood for, and if you are writing about new surgical techniques, is such knowledge necessary? Even where capitals are still employed, it is often enough to know that PTFE is a non-stick substance without knowing that it stands for poly-tetra-fluoro-ethylene.

But other abbreviations need to be spelt out. Writing about problems at Sellafield, one might need 'The United Kingdom Atomic Energy Authority (UKAEA) has authorised the decommissioning of the reprocessing plant'. And, because people often ignore material enclosed in brackets, it might be better to write 'The United Kingdom Atomic Energy Authority, known as UKAEA,....' Again depending on your audience, you might have not only to spell out 'LTE' as 'local thermal equilibrium' but also to explain what it means and why it is an important concept for the work you are describing.

Numbers and equations are another form of scientific jargon; so too are the units used to qualify them. Scientists use very large and very small numbers with relative ease. However, to say that temperatures need to reach 10^7 K at its centre for a star to start shining, or that an electron weighs 9.11×10^{-31} kilograms, is not helpful to the general public. For the first example, 'ten million degrees' will get the point across; for the second, one might have to resort to rounding up and saying 'you need one thousand billion billion billion electrons to weigh just one kilogram', adding 'that is one with thirty zeros after it' to make your point even clearer. But any big figures can put some people off: the distance to the Moon might be better described as a five-year walk.

Most specialist journals prefer equations to read as though they are part of a sentence. If you bear in mind that 'equals' and 'is proportional to' are verbs, it is easy to include most equations in a sentence. However, beware of mathematics which does not contain these symbols: you can tell if the sentence is complete by reading it out loud. For example: 'Therefore $a(b - n) = x$' is a sentence, but 'Therefore $a(b - n)$' is not.

Stephen Hawking was warned that he would lose half his readership for every equation he used in his *Brief History of Time*. So, when writing for a non-specialist audience, you should use words to spell out what, in a scientific paper, would be an equation. 'Global temperatures increase as levels of carbon dioxide in the atmosphere go up' will do. Beware of being too literal, unless it really is essential. 'As you go down into Jupiter's atmosphere, the concentration of methane *increases exponentially*' may be a more literal translation of the equation you have in mind, but your purpose may be better fulfilled by saying that it 'increases rapidly', because more people will understand what you mean. And even if every scientist knows what T and P and m are, write temperature, pressure and mass when you write for the public.

3.2.3 Verbs, Subjects and Objects

Verbs come in a variety of tenses. To tell a straightforward story, choose a tense and stick to it. You may decide to write your story in the present tense, particularly if you are writing about an ongoing project:

> The European Space Agency's remote sensing satellite ERS-1 *provides* atmospheric physicists in the UK with a stream of high-quality data. But continued British success in analysing these data *requires* more funding than the government *is willing* to commit to the project. Nevertheless, project leaders *remain* confident that sufficient funds *will become* available shortly.

Put into the past tense, however, the piece would read:

> The European Space Agency's remote sensing satellite ERS-1 *provided* atmospheric physicists in the UK with a stream of high-quality data. But continued British success in analysing these data *required* more funding than the government *was willing* to commit to the project. Nonetheless project leaders *remained* confident that sufficient funds *would become* available shortly.

Tenses relate events in time, so the logic of the story dictates which tense you choose for which verb. But some careful cheating can sometimes work well: if you write about past events as though they were happening in the present, your text will seem more immediate and engaging.

Verbs must agree with their subject, which is not necessarily the nearest noun to them. For example, in the sentence: 'Our last *set* of *results show* that the cross-section in this region varies smoothly', the verb 'show' is incorrect because the subject of the sentence is 'set', which is singular, not the plural 'results'. The verb should be 'shows'. Remember that the word 'data' is always plural.

The present participle offers plenty of scope for ambiguity. Formed by adding 'ing' to the verb stem, it is often used to qualify the subject of a sentence. But present participles are often used incorrectly: 'After *filtering*, we added the clear solution to our test tube.' This is at best ambiguous, since the subject of the sentence is 'we' and so it sounds as if 'we' were filtered rather than 'the solution'.

Verbs also come in two voices—active ('the man bit the dog') and passive ('the dog was bitten by the man'). In the twentieth century, the passive voice has become the standard in professional scientific communications. This may be because it is felt to convey a greater degree of objectivity. 'The data were analysed using a least-squares fit procedure' makes no mention of the scientist who obtained the data or carried out the analysis, and may appear more objective than 'We analysed the data using a least-squares fit procedure'. Using the passive voice certainly produces a far less personal construction.

But science is about people, and for popular consumption continuous use of the sterile, passive voice is inappropriate. The passive usually results in a wordier construction, for a start: 'the man bit the dog' is five words against the seven of 'the dog was bitten by the man'; and that may be important where space is an issue, as in newspapers and magazines. And, by leaving out the scientists, the writer is omitting the personal side of the science, the very aspect likely to draw in casual readers.

Above all, the most important single thing to remember about verbs is the one which is most often forgotten: *every sentence must have a verb.*

3.2.4 Pronouns

Pronouns—she, it, you, they, he—can be a great source of confusion and frustration. They stand in for nouns as the subjects and objects of sentences, and that means that the correct pronoun must be chosen for each noun. Normally, that is not a problem, but there are some nouns which present a challenge. Is 'the government' singular or plural? The answer is that it can be either; it is up to the writer to make their choice and then stick to it. If you think 'the government' is singular, the corresponding pronoun is 'it'; if plural, choose 'they'.

Pronouns should be unambiguous. Look at this passage:

> Jupiter, the largest planet in the solar system, is surrounded by four large moons and several smaller ones. The closest of the four, Io, is volcanic, its core stirred continuously by its enormous gravitational field. These volcanoes spew ten tonnes of material a second into a tenuous structure known as the Io plasma torus. This is swept up by its enormous magnetic field into a doughnut of charged particles which ring it at a distance of more than 400,000 kilometres.

The first pronoun '*its* core' seems clear—the 'it' in this case is Io. But what of the second—'*its* enormous gravitational field'? If you knew something about the subject, you might realise that it referred to Jupiter. The normal assumption, however, is that the pronoun refers to the nearest noun to which it could correspond—in this case, Io again. By the time we get to the last 'it'—'which ring *it* at a distance...'—'it' could be Jupiter, Io, the Io plasma torus or the magnetic field. The piece could be rewritten:

> Jupiter, the largest planet in the solar system, is surrounded by four large moons and several smaller ones. The closest of the four, Io, is volcanic, its core stirred continuously by Jupiter's enormous gravitational field. Io's volcanoes spew ten tonnes of material a second into a tenuous structure known as the Io plasma torus. This is swept up by the enormous jovian magnetic field into a doughnut

> of charged particles which ring the planet at a distance of more than 400,000 kilometres.

One common mistake with pronouns is to confuse 'it's'—which is short for 'it is'—with 'its' which is the possessive of 'it'. The adjective 'whose' is generally associated with a human subject. 'Of which' should generally be used for animals, plants or objects. So 'Dr Brown, about whose experiments the controversy has arisen, works in New York' is correct. So, formally, is 'the electron, the wavefunction of which we are calculating, is bound by the potential'. But this sounds ugly, and 'the electron, whose wavefunction we are calculating, ...' flows much more easily. One way out of this might be simply to rewrite the phrase as 'we are calculating the wavefunction of the electron, which is bound by the potential'.

3.2.5 Punctuation

Punctuation structures a text by organising the information contained in it into manageable parcels. If the piece is read aloud a rhythm is established which takes into account both the logic of the words and the reader's need to breathe. This is not poetry, so the rhythm is not regular.

On his way home from the 1919 eclipse expedition, Arthur Eddington set out to popularise Einstein's theory of relativity, which his observations had been designed to test. The result was *Space, Time and Gravitation* (Cambridge University Press, 1920). We can look at the way Eddington used various punctuation marks to make his work as accessible as possible to his envisaged audience of interested, educated, but not specialist, readers. Here is some of what he had to say in the chapter entitled 'Fields of Force':

> Although gravitation has been recognised for thousands of years, and its laws formulated with sufficient accuracy for almost all purposes more than 200 years ago, it cannot be said that much progress has been made in explaining the nature or mechanism of this influence. It is said that more than 200 theories of gravitation have been put

> forward; but the most plausible of these have all had the defect that they lead nowhere and admit of no experimental test. Many of them would nowadays be dismissed as too materialistic for our taste—filling space with the hum of machinery—a procedure curiously popular in the nineteenth century. Few would survive the recent discovery that gravitation acts not only on the molecules of matter, but on the undulations of light...
>
> The observer is progressing along a certain track in this world. Now his course need not necessarily be straight. It must be remembered that straight in the four-dimensional world means something more than straight in space; it implies also uniform velocity, since the velocity determines the inclination of the track to the time axis...
>
> Again the observer on the earth is carried round in a circle once a day by the earth's rotation; allowing for steady progress through time, the track in four dimensions is a spiral... Clearly the artificial field of force is associated with curvature of track, and we can lay down the following rule: whenever the observer's track through the four-dimensional world is curved he perceives an artificial field of force.

Sentences begin with a capital letter and end with a full stop, a pause long enough for us to draw breath. Sentences usually contain one main idea and possibly a number of auxiliary ones. Eddington uses sentences of widely varying lengths: the longest is 43 words, in the extracts selected here; the shortest is eight.

The longest sentence, the first, is a gentle historical review of ideas about gravitation. The shortest—'Now his course...'—spits one vital piece of information at us. In the first sentence, Eddington uses commas to separate a piece of subsidiary information from the main thrust of his argument. But he wants you to take it into account with the main sentences, and the comma only gives you the briefest of pauses before continuing. This is a good use of commas and Eddington is rightly sparing with them. For there is a tendency, particularly in scientific writing, where one has always to bear in mind that one speaks only of that which one knows now, and cannot proscribe some future investigator from finding facts contrary to one's current thesis, and thus disproving it, if only temporarily, that is, until the pendulum swings you back into favour once more, to use commas to divide qualifying sub-clause from main sentence, and sub-sub-clause from sub-clause, and so on *ad infinitum*, with the effect that one has, by the time one has got to the end of the, by now, rather complicated, at least in some eyes, sentence, forgotten what on earth the whole business was all about, if it ever were about anything worth remembering in the first place, if you see what we mean.

After the full stop, the next longest pause is provided by the colon. Eddington makes just one use of the colon in the material quoted here—'...we can lay down the following rule: whenever...'. The colon indicates that what follows is a complete idea in its own right. Another common use of the colon is when a list of ideas are being put forward as examples of or reasons for believing a previous idea. An example of this might be:

> We consider that the two following observations indicate that light behaves both as a wave and a particle: interference patterns; and the photo-electric effect.

Notice that as the colon does not finish a sentence, the two 'observations' cited do not begin with a capital letter. And we have introduced the idea of a hierarchy of pauses, by making the semi-colon, the next longest pause, the separator between the two 'observations'. In writing in the popular press, another use of the colon is when dealing with direct quotations, such as 'Professor Peabody said: "We are thrilled at the success of our experiments."' Eddington uses the semi-colon when he wishes to indicate that he is going to amplify the main idea of his sentence. Thus 'allowing for steady progress through time, the track in four dimensions is a spiral' takes us a step further from 'Again the observer on the earth is carried round in a circle by the earth's rotation'. Nowadays, because the injunction against using conjunctions—'and' and 'but'—to start sentences has been lifted, we might not use the semi-colon as frequently as Eddington did. We might start a new sentence, making, for example, 'It is said that more than 200 theories of gravitation have been put forward.

But the most...'. Semi-colons are useful, however, if you'd rather keep things moving; bear in mind, though, that the two parts of a sentence separated by a semi-colon should each contain a verb.

Brackets and dashes perform similar purposes, separating off additional information which could be left out of the sentence without altering its meaning or detracting too much from the overall effect. People tend to ignore information contained in brackets, but one standard use of them is when giving abbreviations, as in 'Fourier transform spectroscopy (F.t.s.)'. A single dash or a pair of dashes gives the reader the impression that you are making an aside or adding an afterthought. Eddington does this twice in the sentence which begins 'Many of them would now be dismissed...'. Here is another example: 'Astronomically speaking, you could say that the whole universe consists of nothing except dirty hydrogen, but—as in politics—it's the dirt that makes it interesting'.

Exclamation marks are frequently used to try and make a piece of writing seem more interesting, or dramatic, than it really is. This is a mistake: either what has been written is arresting or it is not, and the exclamation mark does not alter this. The only use we suggest for the exclamation mark is within direct speech, to indicate the tone of the speaker—as in 'Professor Green exclaimed: "This is an outrage!"'

3.2.6 Gender Bias

Gender bias in writing is an important issue, and nowhere more so than in writing about science, which has the reputation of being a traditionally male preserve. There are many words in the English language that take a masculine gender but really stand for male or female. 'Man' is a case in point, unless a male human is specifically indicated. On its own, 'man' may be replaced by 'humanity'. Attached to other words, it might be replaced by 'person' as in 'spokesperson' instead of 'spokesman', 'chairperson'—or even 'chair'—for 'chairman'.

Another type of gender bias is the assumption that specifically gender-neutral words, like 'scientist' or 'secretary', are either implicitly male or implicitly female. This comes across whenever pronouns are then used: 'Each secretary will be informed that she reports either to the Head of Department, or to his Deputy' is an example. It is perfectly acceptable to use 'they' and its varieties as a singular pronoun, as in: 'the reader can see for themself what we mean'. An alternative approach is to use the plural, where the problem does not arise: 'secretaries will be informed that they report either to...', 'readers can see for themselves...'.

3.2.7 Ambiguity

Ambiguities are very difficult to detect, and it can be very helpful to let someone unfamiliar with your work to look at what you have written. What do you mean when you say 'the loss rate was impressive'? Do you mean it was impressively large or impressively small? Does 'the percentage increased by 5 per cent' mean that 40 per cent becomes 45 per cent, or 42 per cent?

3.2.8 Have Another Look

If you have spent days, or even weeks, on a piece of writing, by the end of it you will feel you know it inside out. You try to read it through to check for the obvious mistakes, but your eyes are not really reading what is on the page. Instead they are interpreting what's there according to the idea you have in your head of what you wanted to put across. If you have the time to do so, we suggest that you put the piece aside for a while, get on with another job, and then come back to it. In that way, not only will you be able to pick up those annoying spelling mistakes or minor slips in construction, but you will also be able to see if what is on the page really does correspond to what you were trying to convey.

You should, if there is time, also try to get a friend or colleague to read your work. Of course, if you are writing for the scientific literature, the article will almost certainly be refereed by fellow scientists in the field. In that case, your colleagues may pick up on errors or ambiguities which can be corrected prior to submitting to your chosen journal. If you are writing for a more popular outlet, choose someone who is similar to your target reader and invite them to be critical.

3.3 Forms and Styles of Writing

3.3.1 Scientific Theses and Papers

Scientific theses and papers are targeted at a specific readership, so you can make fairly big assumptions about what your readers know, and what they will expect from you. Theses are often written in an elaborate, impersonal style with a lot of passive verbs. This may be traditional but it is certainly not obligatory: clarity should be your primary aim. But a thesis must meet the detailed requirements of the examining body to which it is submitted, and it should normally be similar in form and style to other theses submitted for the same qualification.

Scientific papers for publication usually cover a much smaller subject area than a thesis, and so are shorter. You should check the 'Instructions to Authors' that many journals give at the front or back of each issue, to ensure your style of writing meets their requirements. If you can't find any instructions, have a look at the papers published in the journal.

Journals differ widely in style, but for the most part the papers they publish are organised into the sorts of sections which are also appropriate for theses. The following sections describe headings which are typical of a scientific paper or thesis:

Title Page

The title should be brief but unambiguous, and it should give a clear indication of the subject and scope of the work. Below the title, give the names of all the authors and the full postal address of the institution in which the work was done.

Some journals print lists of key words (words used in indexes); you should include in your title any key words that might help people track down your paper. Don't put your name anywhere else on the paper: a title page is easily removed when your paper is sent for anonymous refereeing. The referees may guess who you are, but you don't have to make it easy for them.

Table of Contents

If your paper is long, include a table of contents (this is usually obligatory for theses). This will remind you that you should always number your pages.

Abstract

State the topic, method and main findings, in usually no more than 200 words. A good abstract should summarise the entire paper, not just the most exciting parts. Mention everything new and everything you want people to know. The text of an academic paper for publication should always be double spaced: this also applies to the abstract.

Introduction

What have previous workers done? What did you do? The introduction should begin with a clear statement of the scope of the work, and of why it was done. Refer briefly to other relevant investigations, by yourself or by others, to show how this work relates to earlier research. Mention any new approach, any limitations, and any assumptions upon which your work is based.

Materials and Methods

How did you do it? The materials and methods section should include enough detail to ensure that if the investigation were repeated by someone else, with experience in the same field, they would get similar data. This means that you should

- describe the equipment used and provide illustrations where relevant;
- state the conditions of the experiment and the procedure, with any precautions necessary to ensure accuracy and safety;
- write the stages in any new procedure in the right order and describe in detail any new methods;
- if appropriate, refer to preliminary experiments and to any consequent changes in method, and also describe your controls;

- include information on the purity and structure of the materials used, and on the source of the material and the method of preparation.

Theory and Calculations

In a thesis, the theory section should give the basic principles on which your research is based, refer to previous work, and indicate your overall familiarity with your subject area. This is generally unnecessary for papers: only give newly derived theoretical results, unless the purpose of your work is to demonstrate a novel approach to a subject. You should give enough detail so that your calculations may be repeated. You may wish to give details of computational techniques, particularly if they are new and enable previously impossible calculations to be carried out. You can include computer programs as appendices.

Results

What did you find? The results section should provide a factual statement of what you observed, supported by any statistics, tables or graphs derived from your analysis of the data recorded during your investigation. You could include original raw data in tables in an appendix. The tables in the Results section should be summaries.

Describe successful experiments in detail; mention briefly the unsuccessful experiments and wrong turnings which are part of every investigation. Present your results in a logical order (which is not necessarily the order in which you did the work).

Discussion

Here you present your interpretation of your results. The Discussion should be an objective consideration of the results given in the previous section and should lead naturally to your main conclusions.

Refer to any further light cast upon the problems raised in the Introduction, and say how your work fits in with previous investigations. Claims presented here should be based on your results and their relationship to previous work. Information from other people's work should always be followed by a reference. Check against the original source—not against a quotation or citation—before quoting or citing another scientist's work.

Conclusions

A conclusion is a summary of the main findings and their bearing on the problem tackled in the research. Your results may not be sufficient to conclude a particular debate, but they may make a contribution to its resolution and indicate what future work needs to be done. Such points should be made in the conclusion.

Acknowledgements

In this section refer to the source of finance, and to anyone who helped either in the work—with materials, assistance or advice—or in the preparation of the thesis or paper. A statement may be required, for example by a company or government department, to indicate that the views expressed are not necessarily those of your institution. Always follow the house rules in such a case. Do not use copyright material unless you have written permission. Always acknowledge the source of copyright material.

References

Here you give details of all the references you cite. The house styles of different journals differ enormously, and it is essential that you follow the journal's guidelines on references down to the last comma. All the references cited in the text should be included in the list of references, but no others.

The text of an academic paper for publication should always be double spaced: this also applies to the references. When you quote from someone else's work the quotation must be accurate, and you should give the source of the quotation. Check the accuracy of all references, including the spelling of proper names, since a reference is both an acknowledgement of the work of others and a source of further information for the reader.

3.3.2 Reports

Big organisations which have their expertise spread between separate divisions need good internal communications. You may find yourself working as a research scientist in a government laboratory or commercial company. Your work may involve research into particular problems or be concerned with product development. The results of your investigations need to be conveyed to departments which will then act on your findings. Your readers are busy people and time is money: brevity is therefore important. Millions of pounds, or even lives, may be at stake, so clarity is essential.

Although reports and theses are similar in many ways, there is a difference between them: a report from the materials analysis laboratory to the sales team manager has to communicate technical information to a non-specialist reader. This means that some of the techniques for writing popular articles also apply to reports.

Some companies have house rules which you must follow when you prepare internal reports: ask the people who commissioned the report if they have a preferred style or format. Again, some general rules apply. The front cover of a report should include the title in full, the name of the author, and the organization responsible for the report, with its full postal address. Check to see whether any special notices are required by your company or sponsor.

It will help you to organise your report if you follow a set of headings such as:

- *Purpose.* Who commissioned the report, and why? What are your terms of reference? What constraints and conditions apply?

- *Procedure.* What did you do? Why? Although the amount of detail required for a report is much less than for a thesis, the basic principle is still that you should give enough so that another competent scientist could repeat and check your work. For example, while you would not give the mathematical theory behind the statistical test you used, you would give its name: results analysed using a least squares fit might give a different picture when a Chi-squared test is used.

- *Results.* What did you find? Summarize your results, making use of appropriate charts, graphs and tables, so that the essential findings are clear. Raw data should not be given, but you might mention where it could be found in case it is needed in the future.
- *Summary.* The summary should be a statement of your main findings and the bearing they have on the problem you were investigating or the new product you are developing.
- *Recommendations.* What should the commissioners do next? You should indicate what actions flow from the results of your work in as clear and concise a manner as possible. Remember, however, that those with the competence to act on your findings may be working within constraints that make your recommendations hard to follow. Nonetheless, these additional constraints are not your concern; it is up to the decision-makers to weigh up the various options and competing claims for resources and attention.

3.3.3 Letters and Memoranda

Letters

Every letter is an advertisement for you and your institution. You should therefore take care over the appearance, layout and content of each one.

The basic requirements of letter-writing are the same as for other communications. You must know what you want to say: then, bearing in mind the reader's point of view and likely reaction, you must convey this message clearly, concisely and courteously. Write your letters in everyday English: phrases such as 'I refer, Sir, to yours of the 9th inst.' and 'I remain your obedient servant' are no longer appropriate.

Business letters have a particular form, shown in figure 3.1. Notice these points:

- The recipient's address is at the top on the left hand side. The address is given in full, and it is not punctuated. The lines in the address are aligned to the left; this is also how the address should look on the envelope. The sender's address is at the top of the page on the right (if you are using headed paper, it probably already carries the full address). The date is given in full.
- The letter has a heading, to tell the recipient what it is about. The first sentence reminds the recipient of the correspondence to date.

Vernon McConnell
SysTech Inc.
4505 Northside Drive
Washington DC 40871
USA

Data Services Division
Hillside University
London SW20 9JG

14 October 1998

Dear Dr McConnell

Re: software update

Thank you for your letter of 3 October 1998 in which you invite us to order the latest update to the SysTech programme.

Unfortunately this particular development is not one which we feel would be useful to us, so I shall not be ordering on this occasion. However, I hope you will continue to keep us informed of the progress of your package.

Yours sincerely

Jenny Orton
Systems assistant

Figure 3.1: *Typical format for a business letter.*

- The letter is addressed to a named person, so it ends 'Yours sincerely'. Were it addressed to 'Dear Manager' or 'Dear Madam' it would end with 'Yours faithfully'.
- The writer has given her job title after her name, so that the recipient knows who they are dealing with.
- Any enclosures would be listed below the sender's name under a heading 'Enclosures'.

Memoranda

These are internal letters. Like letters, they are also an advertisement for you and for your department. They should be as short as possible, but contain all the essential information.

Write letters and memoranda on unlined paper and, if possible, type them. Keep a copy as a record. Every letter or memo will probably end up in a filing system, so where possible, it should deal with only one subject.

3.3.4 Essays

An essay can have two purposes. The first is to describe a field or an idea so that the reader understands what the writer wanted to communicate. The second is to describe a field or an idea so that the reader (usually a tutor) can see that the writer (usually a student) understands what he or she wanted to communicate. It is sometimes difficult to explain yourself when you know that the person who will read your essay knows more about the subject than you do; however, whether your reader knows more or less than you, clear writing and an organised structure are essential.

It is very unlikely that you will ever have to write an essay outside an academic environment. However, prospective employers will sometimes ask you to write about why you want the job or to state your position on a particular issue, in which case essay-writing skills can be useful.

An academic essay usually presents an argument or case, or reviews a subject. It should consider only one issue, so be clear about your subject area. Take for example the title: *The Electron.* This is rather vague: is your essay supposed to be about quantum mechanics or the history of physics? Ask your tutor. If the choice is up to you, give yourself a subtitle: *The Electron: a case study in wave–particle duality;* or *The Electron: the rise and fall of a particle.* Now that you know what your subject is, make a plan.

The most important function of a plan is that it forces you first to think, and then to organise, your thoughts. Try starting with a summary. What is the purpose of your essay? What are your arguments? What is your conclusion?

Then write an outline. To achieve a logical structure, write a series of headings and subheadings. (If you phrase these headings as questions, you can turn the outline into an essay by answering the questions.) Then study your headings. Are they all relevant, or have you included some superfluous sidetracks? Is there anything missing, or have you included something that you don't fully understand? You may have to do some more reading.

Once you are sure that your outline is complete and that you understand everything in it, sit down to write the essay. Try to write the first draft in a single sitting: that way you have a better chance of producing something coherent. Then put your work away for a while, and then have another look. Have you achieved your purpose? Are your arguments clear? Is your conclusion sound? You may have to rewrite several times before you get it right, but the more thinking you do between drafts, the more likely you are to succeed.

3.3.5 Feature Articles

The feature article is a format that enables you to write at some length about a piece of science. You have room to make a coherent argument, rather than focusing narrowly on just the latest results. You may have room for some of the history of your subject, and should certainly have enough space to make the connections between the particular work you are writing about and the wider subject, or between the science and issues of general interest to society as a whole. And you may be surprised at the range of publications for which a science feature may be appropriate. Why not an article about some of the great female scientists of our time for one of the magazines catering for teenage girls? Why shouldn't *Country Life* carry something on the chemistry of river pollution? Whether you are aiming at the national newspapers, specialist magazines or a more general-interest publication, you should take the time to read a few issues of your chosen outlet. Each paper or magazine will have its own particular style, which you should follow as far as possible. More importantly, it will give you an idea of what the readership is like and help you pitch your article at the right level. It will also help you to make the connection between your subject and what your chosen publication considers is of interest to its readers.

Structuring your Story

A well-chosen structure can help you to organise your material so that it carries the reader through the article. Sometimes your material will suggest its own structure; on other occasions you may find it helpful to try one of the following:

- *A chronological approach.* The material you are presenting may have come from several different laboratories and have been acquired over a decade or more. Following the emergence of the individual results one by one gives your piece a logical structure. But be warned: your readers have only so much patience and you should give them at least a taste of what the end-result of your story is at the outset. Then they can choose whether or not to follow your tale through to its conclusion.
- *Watching the detectives.* The detective story has long been one of the most popular literary genres and science often fits beautifully into this mode. If your story has a lot of experiments or observations, particularly if they are carried out in exotic locations—searching for neutrinos miles underground or for volcanic shock waves on the slopes of an erupting Mount Etna—the blood and guts style of Mickey Spillane may suit you. Piecing together scant but significant scraps of information to achieve the breakthrough lends itself to Poirot's 'little grey cells' approach. And so as not to make the scientists look too smart, the dead ends and wrong conclusions should not be overlooked.

- *Everyday science.* Cooking is chemistry, tennis is ballistics and car maintenance is engineering. Features about the science in daily life, particularly if they can show how a knowledge of the underlying science can improve performance, have a ready-made point of contact with a general readership. If you want to write a general piece on catastrophe theory in aerodynamics, for example, you might introduce it with an account of a hang-gliding accident.
- *The scientist of the moment.* There are occasions when the individual scientist is the real attraction. *New Scientist* took the election of Sir Michael Atiyah as President of the Royal Society as the occasion to run an extended feature on him and his views. Even if the players are not the centre of the story, it can help a feature article to make them more than just the pair of hands holding the test-tube.
- *Controversial questions.* Since science impacts on almost every aspect of our lives, new developments and discoveries throw up issues to be tackled by society. A feature on some controversial effects of science gives the writer a two-fold task: to present the science as accurately as possible, and to outline and comment on the possible impact. This is probably the most difficult type of article to write, and it is often better to claim less, rather than more, and to tone down your prose rather than exaggerate. Finally, try not to make enemies unless you really mean to.

Getting Started

If you do not interest your reader in the opening paragraphs of your article, the rest will not matter; it simply won't be read. But the snappy introduction is often the hardest part of your whole article. So you should not wait for inspiration before writing anything. It may be that the main body of your text will naturally suggest a good opening. If not, here are one or two suggestions for that all important hook with which to catch your reader's interest from the outset.

You could try asking them a question, preferably one that they really care about. As a way into the effect of car exhausts and industry on local climates, try:

> Is it really true that the weekend is always colder than the days we have to work on? That's the claim in a letter to this week's *Nature*.

Or try something that at first looks surprising:

> If you're a shock wave on Jupiter, it's a hundred times quicker to travel to the centre of the planet and back than to ripple a few kilometres across the surface.

Once more, don't forget the personal touch. If you were writing about a new technique in medical science:

> For Mary Mason, that grey autumn day meant just one more trip to the hospital; more pain, more disappointment was all she expected. For Dr Steve Pullit, however, Thursday 7 November would be a day to remember. For the first time, his revolutionary new treatment would be tried out on a human being.

A Good End

The question of whether or not you have reached the end really depends on whether, given the restrictions of space and the time you have available to write your article, you have said all you want to say. You should have tied up as many of the loose ends as you can, and you may wish to remind your readers of just where you started, and thus how far you have travelled together. Most importantly, try not to think of your ending as something entirely separate from the rest of your article. It should, quite simply, be the last thing you want to say.

Creating an Image

An important aspect of a feature article is that you have some room for the kind of description that paints a mental picture for your reader. For example:

> We are pitched on top of an extinct Hawaiian volcano. At fourteen thousand feet, the thin air leaves you breathless,

> the dryness leaves you hoarse and the cold leaves you numb. Harsh conditions for astronomers, but perfect for astronomy.

The same story in the scientific literature might have read:

> Observations from Mauna Kea were carried out under the following conditions: pressure 600 mbar; relative humidity 5%; temperature –23 °C.

Analogies can be powerful allies. Look at this article by Phil Walker and Nina Hall in *New Scientist* (11 January 1992, pp 34ff), which sets out to explain how the structure of the atomic nucleus may be studied by examining different spin states:

> Suppose someone gave you an egg and asked you to describe exactly what was inside. You could crack it open and take a look. But that might destroy the structure. A better way might be to give the egg a spin. A hard-boiled egg, for example, spins differently from a raw one.

The egg analogy is used throughout for the nucleus, and its simplicity and familiarity make the article understandable and enjoyable. Analogies may be used in an extended form, as in this example, or as short metaphors—'the Orion Molecular Cloud, a vast labour ward for stars'—or similes—'the spin of the electron, like a light switch, has just two states, up or down'.

Even the driest of subjects can be made more palatable by a little imaginative description. See what it can do for a surface physics experiment:

> The weeks of patient building were at an end. Had the glassblower done her job, we wondered? Dr Pringle threw the switch. The vacuum pump chattered reassuringly; no shrieks, no high-pitched whine—no leaks. Time to fire up the laser. Bullets of red light, millions of watts of power, spat out at the target, a billion times a second. Only by subjecting the metal sample to such torture could we find out what was poisoning it, what made it fail as a catalyst.

3.3.6 Book Reviews

Above all, book reviews are a service to other readers, letting them know what books are available, and how good they are. This should be uppermost in your mind when you write a review. Since reviews are often short pieces, it is often tempting to write them after just a cursory glance through the book or—even worse—by skimming the publisher's notes on the cover. But that lets down your readers and may invite problems in the future, should your opinions ever be challenged.

Be sure to look at the style of reviews in the publication for which you are writing. In general, your review should answer some simple questions: what did the author(s) set out to do? Did they do it well? Has it been done better already? Was what they tried to do worthwhile or necessary in the first place? You may wish to give a flavour of the book, quoting a short passage which you feel to be particularly apt or misplaced. Your review should always aim for balance. If you think a book really is nothing more than a waste of trees, tell the reviews editor and suggest the best way to kill it off is to ignore it. At the other end of the spectrum, no book is 'absolutely the last word' or 'a perfect masterpiece'. Time and further work invariably make the 'absolute' relative and the 'perfect' flawed and dated.

The reviews editor will probably have chosen you because they know you to be expert in the area covered by the book you are to review, or because someone else has suggested you. In areas of scientific controversy, you may be tempted to use the review simply as a means of putting your own point of view on the subject. Be wary of this, unless the reviews editor has particularly asked you for an 'opinion piece'.

If you really think that the views expressed in the book are so outrageous that they have to be rebutted, you would do well to contact the reviews editor and ask if you can have the space to answer them. They might suggest you use the publication of the book to write a feature article outlining your views on the subject. Otherwise, you should use a book review to summarise the material in the book, even if you disagree with aspects of it.

3.3.7 News Articles

Although a working scientist or student will rarely have to produce a news article, you can learn a lot about good writing in general from learning how to write news. News articles are short. You may have as few as 200 words in which to tell a rapidly developing story, so every word counts. News articles are written to deadlines; you rarely get a second chance to get the story right. For these reasons, news writing requires tight, disciplined writing.

A news article is based around what journalists call the 'Five Ws': *Who, What, Where, When* and *Why*. A science news article—if there is enough room—may also include *How*. News articles are not only written to deadlines, but they are also often read at a breathless pace. Articles are competing with each other for the attention of the reader. That means the introduction is vitally important. Include as many of the Five Ws as possible in the first couple of sentences. It should be possible for a reader to read nothing but the introduction—the first 40 or 50 words at most—and still have the whole story. The rest of your article is then used to recapitulate, to bring in further details to flesh out your Five Ws, and to introduce less important topics or issues.

An editor will cut a news article from the bottom, and a busy reader may not have time to get to the end. This means that the important information should come as close to the top of the story as possible. It also means that you should not spend too long agonizing over how to tie up your article with a snappy ending: the chances are that your precious finish will be cut.

Look at one of the better tabloid papers and—without paying too much attention to the actual subject and content of the article—analyse how the article is written. In a good news article the sentences are short and active. The average sentence length is around fifteen words.

More than half the words in the article will be verbs and their attendant subject and object nouns and pronouns. Adjectives, which qualify the nouns, are used sparingly; adverbs, modifying the verbs, are hardly used at all. The verbs will almost all be in the active voice. The wordier passive is used only for the less important parts of the story.

Here is an example of how a paper, submitted to an astronomical journal by scientists working at the University of Wales, might be turned into a news story. The abstract reads:

> The correlation of the extended period of biological mass extinctions around the K/T boundary with extraterrestrial amino acids in the sediment record constitutes strong evidence of a cometary cause. While the fact that the dinosaurs' extinction coincided with the Chicxulub cratering and iridium-rich sediments suggests a chance asteroidal or cometary impact, the enhanced input of extraterrestrial matter over 10^5 yr supports the hypothesis of a Jupiter-associated giant comet, fragmented into a multitude of pieces, as demonstrated by Comet Shoemaker-Levy 9, and perturbed into Earth-crossing orbits. Copious amounts of dust were released also, enhancing the dust abundance in the Solar System by several orders of magnitude. The shroud of dust accreted in the Earth's upper atmosphere can be sufficient to impose climatic stresses and cause extinctions of species over a protracted period of 10^5 yr.

The scientific paper then goes on to give the background to the work, describe the precautions the scientists took to ensure the validity of their results, give the results and draw the conclusions. The authors also give the following timetable:

> *65.05 million years before present:* giant comet's close encounter with Jupiter, fragmenting into 1,000 subcomets. Earth picks up organic dust shroud, producing ice-age conditions.
>
> *65.00 million years before present:* comet of mass 3×10^{18} g collides with Earth—forms Chicxulub crater—exploding debris dust accreted by Earth over a few 1,000 years.
>
> *64.95 million years before present:* dust clears, climate warms.

The news story might begin:

> A shroud of organic dust, from a massive comet broken up and flung in our direction by the giant planet Jupiter, doomed the dinosaurs to an icy death. One piece of the comet, weighing 3 million million tonnes, plummeted into the Gulf of Mexico, gouging out an enormous crater. Together, these events shut out the Sun, sending the Earth's climate into an ice age that lasted a hundred thousand years. At least, that's the scenario put forward by astronomers from the University of Wales.
>
> The Welsh team base their theory on increased levels of amino acids—the building blocks of protein—found in mud dating from 65 millions years ago, when dinosaurs last walked the Earth. They claim these amino acids came from space.
>
> The discovery of the recent comet, Shoemaker-Levy 9, shows that Jupiter can smash big comets into several smaller ones. Such break-ups also make a lot of dust, necessary for the astronomers' theory.

In fewer than 160 words, readers have been given the whole story, as put forward by the Welsh scientists. All the five W's and a simple 'How' are included. The news editor reasoned that the death of the dinosaurs was an ever-popular peg on which to hang the story. They also relied on readers being familiar with Comet Shoemaker-Levy 9 (an assumption also made by the authors of the paper), the broken comet which crashed into Jupiter in 1994.

The sentences are generally short: the longest is 28 words, half the length of the longest in the abstract. All the verbs except one are in the active voice. Technical terms go: 'the K/T boundary' is omitted; 'iridium-rich sediments' turn to 'mud'. 'Amino acids' are qualified by 'the building block of protein' on the grounds that, in this diet-conscious age, everyone knows what protein is.

3.3.8 Press Releases

At least some of the science stories which appear in newspapers and magazines, or on the television and radio, result from press releases—brief, stylised reports which are widely circulated. Your institution or company may employ professional press officers to write its press releases, but you may still have to write one yourself. While some reporters reject all press releases as junk mail, others are selective and know a piece of genuine news when they see it.

It pays to liaise with your information office, to supply it with newsworthy stories, and to make sure that it is sending interesting press releases to the appropriate publications.

Good press releases are written in the style of a news story. Figure 3.2 shows one that was sent out by the Physics and Astronomy Department at University College London.

REAGAN HONOURS BRITISH ASTRONOMERS

N.B. WHITE HOUSE EMBARGO—8 pm (20.00 hrs GMT) Thursday, November 10, 1988

British astronomers are to be honoured by US President Ronald Reagan, at a ceremony in the White House on Thursday, November 10, for their part in an international collaboration which has produced a spectacular stream of new discoveries over the past decade.

The International Ultraviolet Explorer satellite (IUE), launched in January 1978, has been chosen from among more than 500 applicants to be one of ten picked for the 1988 Presidential Award for Design Excellence, an honour given only once every four years.

President Reagan will give this coveted award to Professor Robert Wilson, head of the Department of Physics and Astronomy at University College London (UCL), who is representing British scientists involved in the international team responsible for the satellite.

Wilson will explain that IUE is one of the most successful scientific satellites ever launched, saying: 'Our International Ultraviolet Explorer is operating 24 hours a day, and has made thousands of exciting discoveries'.

Alongside Wilson in the White House will be the Administrator of NASA, the American space agency, and the Director General of the European Space Agency (ESA).

Wilson comments: 'This is the highest compliment to the intellectual initiative of groups in this country, who developed the original proposal and concept on which the whole project is based. IUE is an astronomical observatory in space with a telescope that can point to any object and measure its ultraviolet spectrum.'

Contact: Steve Miller, Press Officer (0171) 387 7050 ext 3490

Figure 3.2: *Example of a press release.*

The press release reads like a short news story. It could be published almost 'as is' by a lazy journalist. The text clearly contains the name of the scientist concerned and his affiliation. It also contains a contact number. This is important, because an interested journalist can follow up the press release to get more information from the scientist concerned.

UCL thought that the Presidential Award was of national importance, so this material was sent to science correspondents in both the written and the broadcast media. It is important to target a press release correctly. The announcement of a new scientific course or degree should go to the education correspondents rather than the science reporters, for instance. A particularly local event, like a departmental open day, could find a slot in the local freesheet or radio station.

The *Daily Telegraph* was one of a number of publications that picked up the story. Their version is shown in figure 3.3.

Satellite's designer is honoured by Reagan

By Adrian Berry, Science Correspondent

A BRITISH astronomer was given an unprecedented honour by President Reagan in the White House last night for designing a scientific space satellite that has lasted eight years longer than expected and is still observing the universe.

'Our International Ultraviolet Explorer satellite is operating 24 hours a day, and it has made thousands of exciting discoveries,' said Prof Robert Wilson, head of physics and astronomy at University College London.

For this achievement, Mr Reagan gave him the 1988 Presidential Award for Design Excellence, the first time a British astronomer has been so honoured.

The satellite was launched with the help of NASA in 1978 after the European Space Organisation had decided it was too expensive.

It was the first astronomical satellite ever to be launched into high orbit, 22,300 miles above the Earth.

Being so far from Earth, it can make observations round the clock unlike a satellite in low orbit whose view is obscured by the Earth itself.

'It has so far made 68,000 observations of different objects in the sky,' said Dr Allan Willis, Prof Wilson's assistant.

Figure 3.3: *How a newspaper used the press release.*

This is an example of how a press release is used by the press. Note that the personal side of the science has been emphasised: the press release focused on Professor Wilson, and included a photograph of him which the newspaper printed. The picture considerably increased the story's weight on the page.

3.3.9 Media Proposals

An increasing amount of scientific information is finding its way into print and on to the airwaves. Most of it has got there because an enterprising scientist has found a good story and presented it in an appropriate form to the appropriate journalist. Many more stories never see the light of day either because nobody bothers to contact a journalist, or because the story is presented to a journalist in an inappropriate manner. In this section we look at ways of proposing ideas for different media.

Books

Knowing a lot about a subject doesn't necessarily make you the ideal author. Writing a book requires a huge investment of time, and you should ask yourself if you really can afford, say, six months out from your other duties. Never write a book simply because you feel you ought to have written one, or just because it will help your promotion.

Remember, too, that you must look at the situation from the standpoint of a prospective publisher. Does your chosen area already have a good supply of well-written, up-to-date books? How many people do you think will read yours? Realistically, how quickly can you write the book? This is very important if you want to tie in with a particular event.

If you decide you still have something to say—something which requires a book to say it—look for a publisher who already publishes books like yours. If you are a first-time author, you may obtain a great deal of help and advice from your publisher, if you have chosen well. Write a short covering letter, and attach to it your proposal.

Your proposal should include the following information:

- *Who you are:* a brief CV, a list of past publications and details of any relevant experience.
- *What your book is about:* a working title, and a 300–500 words description, plus a list of chapter headings, with a summary of each.
- *Your reasons for writing the book:* Why? Why now? Why me? What else is around and why isn't it enough?
- *Your target readership:* who do you think will read your book, and why? (Don't say 'the general interested reader'. As any publisher will tell you, there is no such thing.)

You then need to describe the book in greater detail (length, format, illustrations) and to provide a realistic date for submitting the manuscript. Say whether you have started writing and, if so, whether any draft chapters are available for inspection. Once your proposal is accepted your editor will help you with the formal details, and will look after you during the writing period.

Magazine and Newspaper Articles

When you write a book, you choose your own readership and hope that it exists. Magazines and newspapers already have well-defined readerships, so your first task is to find out what that is. Unless you've read every edition of a particular publication, you can only guess at what subjects they've covered recently: if an idea for a feature springs to mind you can be sure that it has also sprung to the minds of several other people, and you may find that *New Scientist* covered your topic only last week.

If you don't want to take a risk, telephone or write to the features editor or science correspondent and ask whether your subject and approach are suitable, and whether or not they've covered the story already. That done, all you can do is write your article and send it to the appropriate person: it is very unusual for an editor to agree to publish an unsolicited article before he or she has even read it.

Radio Programmes

What people forget about radio is that it doesn't have pictures. For a story to be good on radio, it needs to be comprehensible when expressed only in words and sounds. You must also be succinct: if you need to spend minutes describing all the pictures you've had to leave out, your story is not suitable for radio. For both these reasons, the best way to propose a science story for radio is to speak about it rather than write about it. Find out the name of the appropriate person at the radio station, and telephone them. If you are hoping to broadcast your story yourself, this phonecall may act as your audition. Be prepared: know exactly how much time you could effectively fill with your story, and explain it using words you would expect to hear on the radio.

Television Programmes

What people forget about television is that it needs a picture for every moment of every programme. For a story to be good on television it must be comprehensible in a mixture of words, sounds and pictures. Some physics is very difficult to describe in pictures, and sometimes pictures give a false impression of invisible phenomena. You may need to use a great deal of imagination. Because television uses pictures and words, it can convey a great deal of information in a short time. However, this does not mean that it can communicate a great number of ideas in the same period: out of this barrage of information the viewer may glean only one or two ideas from an hour of television. Programme makers cope with this in a way which is rather appropriate for science: they choose one idea, and present it by using lots of examples.

So when you make your proposal, be clear about the ideas you want to convey; then be clear about what examples you would use to express these ideas. Programme makers like to have people in their programmes: which scientists could contribute? Finally, you need to explain what picture will be on the screen at every moment. Aim for a specific slot or series, and find out as much as you can about it. Who watches it? How long are the programmes? How technical are they?

If your story is newsworthy then you must get in touch quickly, so telephone. If you ring a news programme, ask for the science correspondent. If you ring a documentary or magazine programme, ask for the producer or editor. You can find out these people's names by watching the credits at the end of the programme. If your story is not urgent, write your proposal in the form of a script. Try dividing your page into two: in one half write the story in words, and in the other half describe the pictures that will endorse those words. These scripts can be long, so always start with a summary, or send it with a covering letter that outlines your idea.

Some General Points

You should remember that although broadcasters and journalists may be interested in enhancing the public understanding of science and in promoting scholarship, their responsibilities are to their audiences, their employers and their medium, and they know a lot more about all those than you do.

If your idea is rejected it may be not because it was bad, but because you misjudged the level, or there wasn't room for it. Don't let editors hold on to your work unless they give you a firm commitment to publish. If they aren't going to use your suggestion, they will let you know so that you can try elsewhere. Someone else may be pleased to have it. If you decide to try elsewhere, amend your work where necessary—you may find that the editor who rejected it is able to offer some constructive criticism, especially if you ask for advice.

3.4 Layout and Typography

The appearance of a document can be an important factor in whether or not it is read. However valuable the content of your work, your press release will go straight into the journalist's bin if it is typed in elaborate spidery characters, and at the conference no-one will notice your poster if they have to get within six inches of it before they can read the title. Even if your reader is obliged to read your work—the examiners of your thesis or the commissioners of your report—you can still make a good impression by doing them the courtesy of producing a document which looks good and is easy to read.

Nowadays most scientists have access to a computer or word-processor which offers them many of the tools available to professional typesetters and designers. Here are a few guidelines on how to make the most of the opportunities available for good design.

3.4.1 Layout

Layout is the arrangement on the page of text, illustrations and space. There are two factors to take into account when choosing a layout: the practicalities of producing the document, and how it will look

There are many practical questions to bear in mind. What size paper does your printer use? Even if there are options other than A4, your photocopier may only work with A4, so the copies will end up as A4 whatever size you choose. How are you going to hold the pages together? You will need a large margin at the top of the page to accommodate a staple in the top left-hand corner, and a large left-hand margin to accommodate a spiral binding. How will your readers hold your document? You don't want their hands to cover what you've written. Will you print it on both sides of the paper? If so, you may want adjacent pages to be mirror images rather than identical.

How do you want your document to look? Tiny type in broad columns looks intimidating; big type in narrow columns looks trivial. Aim for inviting but authoritative: test out different type sizes in different width columns. A general rule is the smaller the type, the narrower the column. If the type is too small, readers will lose their place when they scan back to start a new line; if it is too large, big ugly spaces will appear between the words.

You can sometimes avoid big word spaces in narrow columns by changing the margin specification. If the beginnings and ends of your lines all line up (as they do in most of this book), your text is 'justified'. Most word-processors have this as default. However, you may be able to change to an unjustified setting such as 'range left' or 'align left' (or right): this means that the word spaces are all the same and one margin is ragged. This gives a lighter, less formal look. The following paragraph shows you what a narrow column looks like when it is ranged left.

> If you have more than one
> column of text on a page, the
> space between the columns should
> be big enough that the reader's
> eye doesn't accidentally cross it,
> but smaller than the margins of
> the page. Experiment to see what
> looks right.

Think about the amount of space you are going to leave blank. White space makes a document look luxurious and inviting, but it also wastes paper. Decide what your priorities are. If the layout you choose produces a document that leaves you with an empty fraction of a page at the end, or at the end of most sections, try fiddling around with margins to distribute that empty space throughout the document.

3.4.2 Typography

The amount of paper you cover has as much to do with typography as it has with layout. The most important factor in your choice of typeface should be legibility. There are two classes of typefaces (or 'fonts'): serif and sans serif. Serif faces—such as the one you are reading now—have strokes on the ends of the characters; sans serif faces do not. Most computers which offer you a choice of typefaces will offer you at least one of each variety: the most common (because they are the most popular) are called Times and Helvetica.

Times is the typeface originally designed for *The Times* newspaper. It is a seriffed face, which means it has small strokes on the end of the characters. Seriffed faces look better in smaller type. Times is compact, and looks smart and formal. Most people are familiar with it, and so they find it easy to read. The main text in this book is set in Times.

This paragraph is set in Helvetica, which is a sans serif font—there are no strokes at the ends of the characters. Bigger type looks good in sans serif characters. Helvetica was designed in Switzerland to be legible throughout the world. It is clear, functional and emphatic, and thus useful for headings. The main headings in this book are in a heavier version of Helvetica.

Once you have chosen your typeface you have to choose a size. Type size is measured either in points (1/72nd inch), or in characters per inch (cpi). Be clear about which you are working with: the bigger the point size, the bigger the character; but the bigger the cpi, the smaller the character. Newspapers generally use 9 point, books 10 or 11 point. The main text in this book is in 10 point type.

You can make smaller type more legible by increasing the space between the lines. The spacing of lines is called the 'feed' or 'leading' (pronounced 'ledding'). Leading is measured from the bottom of one character to the bottom of the character in the line below. The usual leading (and the most likely default option on your word-processor) is two points bigger than your typeface. Most books, including this one, are set in 10/12pt ('10 on 12 point': 10 point type and 12 point leading). You can sometimes get more words to the page by reducing the type size and increasing the leading. Try 9/13pt rather than 10/12pt, for example. If this option is available to you it will be under the heading 'leading' or 'feed', or perhaps 'customise leading/feed'.

3.4.3 Emphasis

Most word-processing packages offer you many different ways of making words stand out from the page. The easiest is to 'display' the words: you set them apart physically from the rest of the text. This is used for quotes, equations and headings. Or you could mark them with a bullet, asterisk or dash. There are also typographical features you can use: try changing the typeface or the size, or using bold, italic, shadowing, or boxes. Unfortunately this wealth of opportunities for adding emphasis often tempts people to overdo it. The word ***<u>**EMPHASIS**</u>*** is in capitals, with asterisks, underlined, bold, and in a different typeface. Although it catches your eye it is difficult to read and looks horrible. Usually one or two distinguishing features are enough.

Avoid using too much of either bold or italic: they will lose their emphasis and the page will look messy. Italic has some specific uses in English: for foreign words (*laissez-faire* and *in vacuo*) and for the titles of publications (*Nature* and *New Scientist*). If these sorts of words crop up often in your document, use something other than italic for emphasis.

3.4.4 Headings

The style of your headings should follow some sort of logical progression. For example, more important headings, such as chapter headings, are usually bigger than less important ones, such as section headings. This progression is known as grading.

You can get your grading right if you classify your headings. Decide which are the main headings (for example, chapters), and classify them as A headings. Then decide which are the next most important headings (for example, sections), and call them B. The subsections would be C, and so on. (Most editors will not tolerate a list of grades that goes beyond E: if you have more you need to reorganise your document.) Then decide on your style for each grade, bearing in mind that the titles will always appear in the order A, B, C, D... (you can miss out grades if that's appropriate, but the ones you use must be in this order). Basic rules for designing graded headings are: bigger type precedes smaller type; bold precedes italic; and displayed text precedes text which is run on (that is, it is on the same line as the text that follows it).

If you number your headings, be consistent. If section 1 is numbered 1.1, 1.2, 1.3, do not number section 2 as 2a, 2b, 2c or 2(i), 2(ii), 2(iii).

3.4.5 Documents for Publication

If your document is in its finished state when it leaves you, you should use layout and typography to make it look as good as you can. However, if your document is going to be published professionally this attention to its appearance may be at best a waste of time and at worst a huge nuisance.

Be sure you know what the publisher intends to do with your work. If it is to be published 'camera ready', the publisher will photograph your pages and reproduce them from the photograph. Your pages should therefore look as good as you can make them. Usually for camera-ready publications the publisher will issue a set of instructions about typography and layout, and you should follow them to the letter. These instructions are to ensure that different authors' contributions fit together, or that a book fits into a series.

However, if your work is going to be typeset, the layout and typography should be as basic as possible. This is usually the case, for example, for submissions to journals. Some authors think that they can save their publisher's time by making their manuscript look as much like the finished product as possible. These authors are mistaken. A beautifully designed manuscript in 10/12pt type is a nightmare for all the people who have to process it: what they really want to see is minimal typography, basic layout, double spacing throughout the document (including the abstract and references) and a clear typeface of a respectable size—at least 12pt if you can. If you want to indicate typographical matters to the copy-editor you may do so—it can help if you mark the A, B, C, ... gradings next to your headings for example—but write in pencil in the margin, and put a ring round anything you write.

3.5 Illustration

By and large, people remember just 10 per cent of what they read, if they are given text alone. But they may remember as much as half if they are presented with images that have been carefully chosen to enhance the meaning of the text, to assist the reader with potentially difficult concepts, or to give a genuine feel for what it is that the scientist is doing. If your work is going to contain illustrations, therefore, choose them with care.

Well chosen illustrations give you an advantage when it comes to trying to get your work published, particularly in magazines and newspapers. They attract the reader's attention and they give the page designer something to work with. An appropriate image may be what gets your article, and not someone else's, into print.

3.5.1 Pictures

There are many branches of science which lend themselves naturally to lots of beautiful images. A whole art genre has grown up around chaos theory; there are many books on astronomy that have a distinctly 'coffee-table' feel about them. And such books have a rightful place in our culture. After all, science is visually beautiful and people can derive great pleasure from well illustrated scientific subjects. But you must make a choice—are you writing captions for a series of great pictures, or are the pictures designed to augment the points you are making in your text? If the latter is the case, then you should use pictures either because they amplify your writing, or because they make a point in a way that words cannot. Do not include pictures simply because you are proud of them. Surplus pictures get in the way, by either distracting the reader or leaving them scanning and rescanning the text to find where they are mentioned.

Just as you should not include surplus pictures, make sure the pictures you do use are not full of surplus information. If you want to illustrate the widget that your group has just made, don't do this with a photograph of the whole laboratory, including the cleaning staff. You may wish to give an idea of scale, however, by showing the whole apparatus, with the widget circled, and then a blow-up of the widget itself. Alternatively, a ruler placed alongside your pride and joy may suffice.

Unless you are doing a 'then and now' comparison, ensure that illustrations are up to date, whether they be of locations, laboratories—or people. (Don't you just cringe when the Open University repeats that programme you made wearing a flowery shirt and flared jeans?) Half the world's human population is female; not everyone is a white Anglo-Saxon. Do, or should, your pictures reflect these facts?

False colour is something which is very popular nowadays, whether it be a NASA simulation of the surface of Venus or temperature contours in a heated crystal. Do be sure that your captions make clear what the colours mean. Temperature is particularly tricky; blue stars are hotter than red, but the popular imagination has blue as a cold colour and red signifying warmth. So try to choose a helpful colour scale for your pictures.

You should ensure your pictures are in keeping with your writing, particularly if the tone of your writing is argumentative or polemical. 'The nuclear industry is cavalier, dirty and unsafe,' you declaim unsuccessfully, beneath a photograph of scientists with concerned expressions and in spotless white coats, handling spent fuel rods with remote handling equipment, behind lead glass screens in a laboratory plastered with warnings from the management of BNFL. And the idea that 'scientific whaling, carried out humanely, is essential' is not advanced by an image of a graceful, gentle beast swinging unceremoniously from the mast while its guts spill onto the deck below, splashing ape-like, guffawing matelots of dubious sobriety. If your photograph shows an unusual view of your object, or your illustration is an unfamiliar projection, it is useful to indicate how this relates to what the reader is used to. This may be done in the caption: 'Here we present a map of the world with the size of the countries scaled to their respective populations' might be useful for an article on population density. Alternatively, the addition of a familiar view of the object, with lines indicating how the view you are showing relates to it, may help.

Labels are important for pointing out specific parts of your picture, especially if subtle features are concerned. But the lines between labels and their respective features should not overlap, if it can be avoided, and you should ask yourself if minor details really need to be pointed out. Would the diagram be clearer if fewer labels were attached?

3.5.2 Graphs

While scientists make great use of graphs in communications between themselves, it should be remembered that they are far from being an everyday aid to understanding. Even within the scientific community, many graphs are inappropriate. Too often, the wrong quantities are shown, or the ranges of the axes are unhelpful. Do you need the origin on both axes, if the range of values on the ordinate runs from 100.0 to 110.0? Three dimensional surfaces are often difficult to visualise, especially if the surface is awkwardly folded. Might a contour plot be clearer?

As far as scientific articles for popular consumption are concerned, the simpler the better. Graphs might be replaced by pie charts. But if the sections on the pie chart are similar in size, a bar chart showing only the tops of the bars might be better, as it is generally easier to judge lengths than angles. Your graph will have to fit across either the full page or a number of columns, so draw it so that it is roughly the right shape.

3.5.3 Illustrations for Publication

If you are supplying illustrations for publication, the following guidelines apply.

Your publication will tell you what sort of pictures they are prepared to publish. The types of picture will probably be specified in the following terms. Any picture which contains only black, white and shades of grey (such as a black and white photograph) is called a 'half-tone'. One with a range of shades of different colours (such as a colour photograph) is called 'full-colour'. A drawing that contains black and one other colour is called 'two-colour': black counts as a colour but white (empty space) does not. A drawing which contains only one colour (and no shades of that colour) is called a 'line drawing', even if some parts of the picture are not made up of lines.

A black and white photograph should be provided as a print, but a colour photograph can normally be printed from a transparency or a colour print. Draw your line drawings in thick pen and much larger than they will appear in print. They can then be reduced photographically and will look much neater than a small picture printed actual size. Test your pictures by reducing them on a photocopier to see if they are still clear. Computer print-outs are rarely appropriate for publication. In the first instance only send copies of pictures: keep your originals until you are sure your article is going to be published.

Further Reading

Friedman S M, Dunwoody S and Rogers C L 1986 *Scientists and Journalists: Reporting Science as News* (New York: Free Press). An insightful study of the relationship between science and the mass media based on analyses from both points of view. Lots of practical advice, and annotated bibliographies.

Huff D 1988 *How to Lie with Statistics* (Harmondsworth: Penguin). Includes a section on how not to lie with graphs.

Kirkman J 1989 *Effective Writing* (London: Chapman and Hall)

Kirkman J 1992 *Good Style* (London: Chapman and Hall)

Lettering and Typography. An Usborne Guide 1987 (London: Usborne Publishing)

Quilliam S and Grove-Stephenson I 1990 *Into Print* (London: BBC Books) Chapter 7

Strunk W Jr and White E B 1979 *The Elements of Style* (London: Macmillan). Now in its third edition after 30 years, *Elements of Style* lays down the law on usage, composition and style. A must for any unsure author.

The Concise Oxford Dictionary (latest edition)

The New Collins Thesaurus

The Oxford Dictionary for Writers and Editors. Very useful for abbreviations, proper names, and new or colloquial words, and words that can be spelt in a variety of ways.

Webster's New Collegiate Dictionary (latest edition)

CHAPTER 4

WORKING WITH THE MEDIA

Jane Gregory and Steve Miller

4.1 Introduction

There is enormous scope for mutual misunderstanding and suspicion between scientists and journalists. In many cases, scientific research is carried out by teams whose results accumulate over a number of years, bringing gradual changes in understanding, and introducing new models of the natural world only after old ones have been tried and tested to the limit. The media are geared up to deal with individual personalities and events, stars and breakthroughs. New developments in science may be no more than a subtle, yet critical, modification of what was known before; understanding their true significance requires a good knowledge of the entire field. Journalists want the whole story in a few soundbites—their readers or viewers are not experts and need the information in a handful of pithy sentences.

What for a scientist is being precise, is, for a journalist, splitting hairs. Necessary scientific qualifications translate as uncertainty or the hedging of bets in the media. To become accepted by the scientific community, new results or theories have to go through the lengthy process of refereeing and publication in the relevant journal, but the media want the latest thing and they want it now. Scientists who talk to the media may be seen as trying to steal a march on their rivals or circumvent the process of peer review.

Despite all the possibilities for tension and hostility between the media and scientists, there is a long and honourable tradition of reporting of science to the general public. The Victorian general interest magazine *The Athenæum* had regular articles on science mixed in with reviews of art and politics. Newspapers have always seen science as a subject of great interest. *The Times* gave extensive coverage in 1919 to Eddington's eclipse expedition to test Einstein's theory of relativity, and the tone was not much different from that of the 1992 coverage of COBE's ripples in the cosmic microwave background.

Science on the radio made real headway during the Second World War: the Government saw it as a useful way to get information to the public about the nutritional value of what foods were available. There are now frequently science features on radio, and there are several series of specialist science and technology programmes. The first science programme on television was *Inventors' Club*, which was broadcast from 1948, and *The Sky at Night* has not missed a month since it started in April 1957. Alongside the long-running documentary series, science also crops up every day in a wide range of news stories and features.

And, in general, when the media cover science they do it well. Journalists are no less professional than the scientists they deal with; all but a few want to report accurately and fairly. As they see it, the job they are doing is to stand in for their audience, asking the questions that the general public want answered. If you are the person on the other side of the notebook or microphone, then understanding a little about the media and following a few simple rules will help you do yourself and your subject justice.

4.2 Contacting ...

If you are going to talk to the media it is important that you have something useful and interesting to say. Your dealings with journalists may either be initiated by yourself (or your institution), or by them. If you are doing the contacting, it may be because you think you have an exciting result which should be given some public attention. But before you make that call or put out that press release, here are a few questions you should ask yourself.

- *Why is my work significant?* Is your latest technique really going to revolutionise medical physics? Or have you just made one among many useful, but rather minor, improvements to magnetic resonance imaging? Have you really solved the riddle of the extinction of the dinosaurs? Or does your experiment simply remove one of the objections to the crashing asteroid theory? Journalists may not be specialists in your chosen field, but they usually have good instincts and can tell if they're being sold a dud. So pitching your approach correctly is important.

- *Why is my work interesting?* One question a journalist will always have in mind is: 'So what?' They will be considering who, outside a small clique of specialists, would want to read or listen to the story you are telling. That means a small step forward in the understanding of a problem that affects many people may be more newsworthy than a revolution in an obscure branch of quantum electrodynamics. That said, the media do often cover stories that fall in the 'curiosity driven, science for science's sake' category.
- *Is now the right time?* This is really two questions. The first is whether you feel that your work is at the point where you want it to become publicly known and available. Would it be better to wait for that third referee's report, even if the first two have been favourable? Is there just one last test we should try out on our robot to make sure it pours the tea into the cup and not all over the carpet?

  The second question is whether there is a convenient topical peg on which to hang your work. That seminal research you did on turbulent wind systems three years ago—and which was largely ignored at the time—acquires a new significance as a result of the catastrophic collapse in a gale of a local suspension bridge, and journalists may be hunting for expertise such as yours.
- *Whom should I contact?* Science stories are generally handled by the science editor or correspondent. This will certainly be your first port of call for a story that is of the 'science for science's sake' variety. But the news desk would be interested if your work had bearing on an outbreak of cholera in Manchester or the appearance of bright lights in the sky above Glasgow. The education correspondent is the person to contact if you have found a way to teach differential calculus to primary school children. And what is of no interest to a general newspaper may be a red hot topic to *Computer Weekly*.

4.3 ... and Being Contacted

Many institutions send out press directories, listing the areas of expertise of their research groups together with the relevant telephone numbers. A journalist working on a particular story may need some comments, help or advice, and may get in touch with you as a result of this. Or you may have allowed yourself to be put on a general database of scientific experts. It may also be that your work has been the subject of a press release put out by a prestigious publication or by your company, and that this has attracted media attention. Or an enterprising television producer may have heard—or got to hear of—some comments you made at a recent conference that they feel are worth following up. You may be contacted by the press for these or a host of other reasons. At this stage, there are some useful questions to ask, of the journalist and of yourself.

- *Why and how have I been contacted?* You should establish early on in any conversation you have with a journalist just what their purpose is in contacting you. Are you even the right person? Have they mistaken your listed expertise in *quantum* control systems for *quality* control systems? Find out, before you say anything or agree to anything you might later regret, just what story they are working on and how you fit into it, if this is not immediately obvious.

  It can also give an insight into what the journalist is after if you ask them how they got your name. Was it through a database or college directory? If someone you respect suggested you as just the right person, you may feel more at ease than if you were the only person in at the time the local press called.

- *Whom am I speaking to?* At the start of any contact with the media you should find out the name of the journalist you are talking to, the name of the publication or programme (including station or channel) they are working for and their position (science editor, programme researcher). Knowing who you are talking to tells you not only about the journalist concerned but also about the wider audience you are addressing. That, in turn, helps you to give useful information at a level which is appropriate. You may also decide that explaining astronomy to the fashion editor of *You and Your Horoscope* is not worth the effort.

- *Do I really want to do this?* It may be flattering to be asked to take part in a local radio discussion on spoon-bending the first time you are asked. But after a while, debating with psychic phenomena enthusiasts can pall. So while we urge scientists to be as co-operative with the media as possible, don't be afraid to say 'no'. Be careful, too, if it is clear that the journalist is trying to write a controversial piece. You may not want to be quoted in print as calling the current leading expert in your field 'a self-important charlatan'. One way round this is to insist, at the start of any conversation, that your comments are 'off the record' and for background information only. You can always go 'on record' once you are sure of the journalist's angle and intentions.
- *Do I have the time?* There is no point agreeing to speak to the media if you really have seven research proposals to finish by the end of the week. Try to get some idea of how long the journalist wants to talk to you for (and add on half as much again). Try also to find out how long the radio programme will be that you are contributing to, and what percentage of it the journalist hopes your material will fill. That also gives you some idea of how long the interview might take—but bear in mind that it can take 20 minutes of conversation to produce 15 seconds of radio.

4.4 Being Interviewed

If you are asked to give an interview to the media, rather than just a few comments over the phone, you can assume that the information you give or the views you express will occupy a significant part of the programme or article that is being put together. But it is rarely as much as you think you deserve. Good journalists try to get balance in controversial articles, and may use a number of 'experts' on both sides of the argument.

A television science correspondent may well recast much of what you tell them into their own words, broadcasting a mere 30-second clip of you talking to camera from a whole afternoon's filming. Whatever the eventual outcome, the clearer you are, the more of your material will see the light of day, and the more likely the journalist is to use you again.

4.4.1 Preparing for an Interview

Unless you face a situation where a journalist complete with camera crew has waylaid you on your way out of the laboratory, you will have an opportunity to prepare in advance for the interview. As far as possible, ensure that it is at a time and place that suits you. If a radio reporter is coming to your office, make sure you won't be interrupted by people knocking at your door or the telephone ringing. Will the camera crew really be able to fit into the laser laboratory? Have you allowed enough time not only to do the interview, but also to get to the newspaper offices?

But—and this is for the supremely confident who want to be truly spontaneous—that may be all the preparation you do. At the other extreme, you want to may research not only your subject, but also the interviewer and every programme made for the series they are currently working on. In that way, you might hope to have a prepared answer for every question they could conceivably ask. Somewhere in between these extremes lies an adequate level of preparation. You should know what the subject of the interview is, and have polished up your knowledge of it, and have a good idea of the target audience. Don't become obsessed with your appearance if you are going to do a TV interview—you don't need to 'dress up'. Find out what other contributors to such programmes have worn, and make sure that you feel comfortable with the way you look.

Colleagues may be able to tell you if their dealings with your prospective interviewer were satisfactory from their point of view. If you are talking to the local radio station, are you going out live, or will your interview be recorded and cut into another programme? Will you get a chance to see a journalist's rendition of your comments before they are published?

4.4.2 The Interview

If there is time beforehand, you might ask the interviewer to run through the questions with you to make sure they are appropriate and cover the ground you think is important. If the interviewer doesn't want to do so—for whatever reason—don't insist. Each journalist has their own way of working, and if they are planning to tear you to pieces, they are not likely to forewarn you. But this rarely happens: it is usually safe to assume that your interviewer is genuinely interested in the answers you have for their questions. Assume they are friendly until they turn really nasty. This means that you should answer the question you are asked as simply and effectively as you can. Don't try to second guess your interviewer, wondering all the time about their hidden agenda. Answer honestly, and steer clear of jargon. Be yourself.

As much as possible, relate your work to what you think your audience will find relevant. Or, at least, give them a point of departure which they recognise: talk about their reflection in a mirror or the way flower petals are arranged before you head off into the mysteries of symmetry.

If the journalist asks a question you do not understand, ask them to repeat it. If you do not know the answer, say so. Don't try to make the journalist look foolish or ignorant. They may simply be putting the questions they think their audience would want answered. If you are not doing a live interview, and you do not like the way you answered the question first time round, ask if you can try it again.

At the end of the interview, you may feel that the interviewer has missed an important point of your work, or that there is something you really wanted to say that hasn't come out yet. If you point this out, it is unlikely that the journalist will not let you say it.

4.4.3 After the Interview

Once a live interview is over, there is nothing you can do about it. If it went well, feel pleased; otherwise, chalk it up to experience. If there is going to be a delay between giving your interview and the broadcasting or publication of the material, make sure the journalist has a number on which they can contact you, should they want a point clarified or expanded. But it is probably not worth your while contacting the journalist again: unless you suddenly remember a really crucial bit of information that you left out, a call to the interviewer will appear to them as if you are checking up on them.

4.5 Broadcasting

In chapter 3 we set out some of the ground rules for producing articles for popular consumption in a range of possible outlets. Many of those rules—avoiding jargon and equations, using simple rather than more complicated words and phrases—apply equally well to broadcasting. There exists a wide literature on television and radio techniques, and a detailed exposition of how to become a David Attenborough or Carl Sagan is beyond the scope of this book. Nevertheless, here are some basic pointers that will help you make the most of what are likely to be rare—and potentially valuable—opportunities to broadcast.

Television and radio carry a wide range of programmes about science, and a programme editor may feel that your work could benefit from a treatment more extensive than just a short interview—perhaps an hour-long programme. That is, they feel it is worth the broadcasting equivalent of a newspaper full-page feature article. Or you may have an idea for a whole series of programmes, the equivalent of writing a book with a number of well-defined chapters. To get anywhere with your idea, you will first have to write a media proposal, as described in chapter 3.

You will also have to make sure that you really do have the time needed to make the programme. When *The Sky at Night* decided to make a programme about the International Ultraviolet Explorer satellite, they required three half-day meetings to discuss the script with University College London's Bob Wilson, and shooting the 20-minute programme took a whole day. You may need to ask for a year off work to make a series of programmes. But while your lecture or seminar may reach an audience of a few hundred at most, television and radio programmes can reach millions. It is up to you to decide whether that is a good use of your time.

4.5.1 Television

The obvious advantage of television is that it is a supremely visual medium. But that can also be a daunting challenge. During every moment of a television programme, the screen must be filled with an image. And it is no use hoping that half an hour watching a scientist talk to camera will do. That may be fine for the handful of scientific superstars, where the chat show or in-depth interview formats work because the audience is interested in the person, rather than in the scientific subject. But, by and large, talking heads make boring television.

That means your subject must be naturally visual, or that you—or the programme maker who has invited you to participate—must be very imaginative. If the budget allows, one obvious ploy is to use a variety of locations. If you cannot afford the fare to Pisa, you could star in a programme on Galileo by dropping various objects off the Blackpool tower, followed by a trip to the observatory at the University of Central Lancashire to look at the moons of Jupiter.

When you propose a programme on the latest thinking in high energy physics, you should have some idea of how you can use computer generated graphics to illustrate the production of exotic particles. Can you use ripples in a water tank to explain gravity waves? In television, you use pictures to make arguments that you would normally write out in words. You use images as explanations. But whether the programme or series results from your proposal or from an invitation to a programme maker, you will not be making it alone. Essentially, you bring the science to the programme. Television is well staffed with experts in visualisation, and producers and editors know how to cut material together to make a coherent and entertaining whole. Camera operators know which shot to use in almost every situation. And if you can clearly explain the science that you want animated to the visual effects artist, you can let them get on with modelling it.

4.5.2 Radio

The obvious disadvantage of radio over both television and print is that there is absolutely no opportunity for using images. Paradoxically, that is what has ensured that radio has survived and prospered. From the standpoint of the audience, listening to the radio is nothing like as limiting as watching the television. You can carry on with the washing up or be driving your car and still take in most of what comes over the airwaves. And because the technology is much simpler—you can make a radio programme with nothing more than a good quality microphone and tape recorder—radio is much less intimidating than TV for the occasional broadcaster.

Even with the advent of cable and satellite stations, there are still many more radio than television outlets. As well as the BBC's five national channels and the World Service, there are innumerable local stations. There may also be local college and hospital radio stations, and these can be very receptive to suggestions as they are likely to be less well supplied with interesting material.

Since there are no pictures to get in the way, radio listeners pay much more attention to the words. Radio is an excellent medium for making an argument or having a discussion about an contentious scientific issue. Do remember, however, that unlike readers of an article in a magazine or newspaper, radio listeners cannot go back over a bit they did not understand first time. Either make your exposition very simple and linear, or give yourself time to recapitulate slightly—to summarise what you have already said—before launching off into the next strand of your argument.

As there are no pictures, sound alone has to create the image you want to convey. Your words will bear the brunt of this task. As with writing, analogy with everyday situations is a great help. But don't forget that radio has developed great expertise in sound effects to help get the message across. If your design for a neural network computer is based on a model of how information is passed through the swarm in a beehive, the murmur of innumerable bees may set the scene for your explanation. Even if you are simply being invited on to someone else's programme, you can still suggest suitable sound effects, just as you would images for a TV programme. And well-chosen music can enormously enhance your broadcast's appeal.

4.6 Coping with the Fallout

Once your work has been given an airing in the media, or your face has appeared in the homes of the nation, you may expect further attention—especially if you came across well or stirred up some controversy. This may range from invitations to take part in more programmes or to write an article expanding on your views, to the opportunity to review *The Theory of Creation, Incorporating General Relativity and Morphic Resonance*, written in purple ball-point on the back of an application to become manager of the England football team.

Some of your colleagues may be pleased that your foray into the public domain brought credit on you and your institution; others may blame you for millions being wiped off your company's value on the stock exchange. Your life may not be changed for ever, but you can expect at least some response. Unless you have decided that your future lies in becoming the next presenter of *Tomorrow's World*, ensure your additional commitments do not make it impossible for you to do the job for which you actually get paid, or the research specified on your grant. And try to see that your new found fame (if any) does not inconvenience your colleagues—they'll still be around when the fuss has long since blown over.

Part 2
COMMUNICATION THEORY AND PRACTICE

CHAPTER 5

BASIC CONCEPTS AND PRINCIPLES OF COMMUNICATION THEORY

Shirley Earl

5.1 Introduction—the Need for Effective Communication

'I always call a spade a spade...'

'It's straightforward—ask the boss.'

We all think we're good communicators. After all, we began communicating the day we were born and by the time we went to college or started work we'd been doing it for something like sixteen years. Things may go wrong occasionally, but we generally manage to salvage the situation. Rarely do we stop to see ourselves as others see us.

But each of the comments above could refer to a situation perceived very differently by someone else:

'Guaranteed to put his foot in it.'

'You must be joking.'

Far too often there is a gap between what we say and what we think we have said, between how we think we handled people and how they feel they have been treated. Relationships and people's physical, mental and emotional well-being can be damaged when these gaps become chasms. A 'breakdown in communication' can be very serious indeed: when relations reach a low ebb relatives can cease speaking, tensions mount on the factory floor, management and trade unions refuse to meet and governments break off diplomatic relations.

This chapter is founded on the idea that problems are liable to occur whenever people communicate, whether as individuals, within groups or on behalf of organisations. Offence may be taken at a remark, instructions can prove difficult to carry out, ambiguous phrasing may cause confusion, and reactions can be coloured by previous experience, suspicion, pressure and stereotyped responses.

Today there are many organisations where hundreds of employees work on one site, and companies which employ thousands of people and have links with suppliers and customers worldwide. Good communications are clearly essential in such large organisations. Equally, however, it is dangerous to assume that a small organisation will be immune from communication problems simply because it is small. The only advantage of being small in this context is that problems may be more easily identifiable and remedied. Problems are just as likely to occur.

For these reasons managers and social and political scientists have devoted much energy to analysing problems caused by poor communication, and to the careful consideration of communication concepts, the development of sound communication theories and the creation of good communication climates and systems. To understand the environment in which we work (and teach and learn) it is important to think about the foundations and skills of communication.

In this chapter I hope to stimulate thought on three basic issues: the modes of communication we use in our working lives, the simple stages in the communication process, and basic theories of communication. More specific techniques of *verbal* communication are covered in chapters 3 (writing) and 6 (speaking). *Non-verbal* communication—everything apart from the actual words—is discussed in section 6.2.2. It is used in organisations, as elsewhere, to modify or augment the verbal message, though on occasions it may be employed to convey meanings that are contradictory to the words used.

5.2 Patterns and Modes of Communication

Communication within every working organisation is both *formal* and *informal.* Formal communication is represented in the patterns and structures of working relations—the flow of information, of authorisations and of instructions up, down and across the hierarchy. It takes the form of reports, briefing documents, minutes, administrative procedures, codes of practice and more. Not all of these forms are written. An oral report is as much a report as a written one.

Informal communication on the other hand is less easily categorised and controlled, but it nonetheless exists and is powerful. Unminuted discussions, chats, notes, telephone conversations and confidential briefings are all examples of informal communication. An organisation's grapevine is part of its informal communication network; its house magazine is not. The former carries unauthorised information and is difficult for senior management to access. It is nonetheless vibrant, or virulent, according to our individual point of view and the content of the moment. The latter is a currently popular medium intended to boost morale, keep all staff abreast of events and cement corporate identity.

The way in which communication flows in an organisation—both formally and informally—has far-reaching consequences. It can provoke reactions from staff, affect their performance and facilitate or inhibit the exchange and change of ideas, attitudes and feelings. Communication can simultaneously reflect the purpose of the organisation, establish its standards and encourage or discourage involvement. Ideally, the structure of an organisation, its climate and its communications complement each other; for example, it is hard to have free flow of information within a rigid authoritarian hierarchy, and an organisation wedded to task-group structures and devolved responsibility will not easily accept new restrictions on access to information.

Hierarchical structures seem to be unavoidable in all but the smallest organisations, with the pyramid the most common hierarchical form—though others are evolving. Experience, expertise, external expectation, and the need to have decisions executed and results monitored, are all pressures towards pecking orders and pyramids. These pressures are reinforced by the division of organisations into specialist departments, all answerable to a more senior coordinator. Each department has its own pyramid structure and communication patterns are ultimately, officially or unofficially, constrained by them.

This gives rise to identifiable communication routes, flowing up, down and across, from shop-floor to boardroom and back, between section managers or hands. Flow also occurs between different levels of separate specialist sections and between the internal organisation and external suppliers or agencies.

Vertical Communication

This term describes communication from top decision makers to people at various lower levels and, correspondingly, the flow of information, ideas and queries from subordinates to people higher up. In a healthy organisation an upward flow of ideas, suggestions, criticisms is just as important as a downward one. If downward communication becomes an avalanche and upward communication weak then the organisation is at risk of explosive frustration, poor morale and declining productivity.

Horizontal Communication

This is also known as lateral communication, and occurs between people on similar levels. It tends to be less inhibited than vertical communication because of the absence of hierarchical, chain-of-command constraints. Horizontal communication occurs at all levels of an organisation and frequently signifies ease and frankness between equals. Weaknesses in horizontal communication often indicate that the organisation is suffering from rivalries and empire building.

Diagonal Communication

This spans levels and specialist departments, and requires co-operation from both sides if it is to function well. Because diagonal communication crosses hierarchical chains-of-command it relies heavily on the sort of co-operation and goodwill which an effective communicator will have been careful to foster informally, by friendly conversations with colleagues, on journeys through departments or at social functions. Geographically separated locations can inhibit diagonal communication.

Internal Communication

This is simply communication within the organisation. It may flow along vertical, horizontal or diagonal routes.

External Communication

This term refers to interaction with outside agencies—suppliers, customers, statutory bodies. Internal communication generally presents fewer cultural problems than external.

One of the most challenging tasks for people in organisations is keeping all the routes of communication open. Each of us is a user and a distributor, and each of us is also a person as well as an employee. When any of these four factors is forgotten problems are likely to arise.

5.3 Getting the Message Across

If your communications are to be accessible you must make the process interactive, by taking other people, and their reactions and environment into account. Because things have to be considered from both your own point of view and that of others, communication specialists tend to speak about two linked facets of communication, those of the transmitter (sender, distributor) and of the receiver (user). With the receiver feeding information back to the sender, the process becomes *circular and reciprocal.* Communication is seen as a continuous flow rather than a simple matter of stimulus and response, a continuum within which one exchange builds on another.

There are six important stages in the circular process, namely forming the idea, encoding it, selecting the route and sending the message, decoding, interpreting the message, and feedback.

Forming the Idea

In this first stage the 'transmitter' decides what to convey. Some messages are formed involuntarily—such as a shout of anger or a cry of pain. Others are the result of a mental process so quick that we are scarcely aware of the thought—the danger on these occasions being that one may 'put one's foot in one's mouth'. Yet others are the controlled result of a careful process of reasoning, for example when deciding whether to insert a penalty clause in a contract.

Encoding

Before any idea or message can be sent it needs encoding. In the examples in the preceding paragraph three different 'codes' are appropriate: a shout, emotional spoken words, and precise written legal terminology. Encoding is putting communication into a recognisable form by choosing an appropriate 'language'. There are many languages of communication, including the spoken word, the written word, number, graphic illustration or icon or symbol, and non-verbal presentational codes which use the body as a transmitter.

The ultimate success or failure of a message depends very much on appropriate encoding. Each 'language' has strengths and weaknesses—speed of the spoken word but difficulties with verification, the verifiability but time-consuming nature of the written word, conciseness but cultural differences in concept of number, conventional and representational dimensions of illustrations (e.g. whether a specific poster is one of a man at work or a man at risk) and ethnic and gender differences in use of the body to convey meaning.

Selecting the Route

Once the appropriate language has been chosen, the sender comes to the third stage of *selecting the right route* (communication specialists use the term 'medium', but a lay translation is 'route'). This should be a conscious choice, dependent on a number of factors. Some of these are time, cost, the need for rapid feedback or simultaneous reception or a written record, the complexity of the message and the relative status of sender and receiver. If instant response is needed then a head round the door and a few spoken words may be required. If the message is complicated and detail is important then a written document is called for, to be studied at the receiver's leisure. If it is near the end of a working week and people are tired then bad news may best be delayed; the same news may be better relayed personally and face-to-face rather than in a written memo or, worse, by e-mail.

With modern communication systems an increasingly refined choice of route becomes available. If we choose computerised datalink, for example, what access codes must we use to improve confidentiality? Would the moving image of a video be sufficiently more powerful to outweigh the comparative cheapness and greater adaptability of a tape-slide presentation? And so on. The point is that to put your message across effectively you must consider carefully the available routes and media and their appropriateness.

Decoding

Whatever route you choose, the message that it carries cannot be absorbed until it has been understood. This is the point at which the receiver rather than the sender becomes an active partner in the communication process—the stage of decoding. Many messages fail to be understood because the sender has chosen a language (in the technical sense discussed above) beyond the receiver's comprehension. As an example, I recall looking with an American friend at the purple and white expanse of a hoarding advertising a well-known brand of cigarette. To me, conditioned to expect a bland government health warning and to associate the image of cut silk with a particular brand of British cigarette, the hoarding was an enticement to smoke. To my friend the words looked stronger than the image. He saw an anti-smoking poster, thundering out the message 'Smoking can damage your health', and reinforcing it with the cut silk image which graphically depicted surgical incision.

The sender of a message has no guarantee that their chosen code will be known to the recipient; what they do at the encoding stage is try to ensure the best chance it will be understood. The decoder does the rest taking into account factors which are similar to those which the encoder considered.

Interpreting the Message

Some people see little distinction between decoding and interpreting. The important thing is that as well as understanding the message, the receiver needs to be able to interpret it correctly—to work out what the words mean as well as what they say. Many messages are capable of more than one interpretation, and these interpretations may be at variance, as in the example of the cigarette hoarding. Sometimes messages are deliberately encoded to carry an apparent and an underlying meaning, and thus to reinforce a point. Technical presentations may really need to be complex, or they may be phrased falsely to impress. Spoken colloquialisms uttered in a tone of sarcasm must be interpreted differently from the same words articulated with sincerity. In addition, the context in which a message is sent affects the way it should be interpreted, a factor which makes particular difficulties for time-lapsed communication.

Feedback

This is the sixth stage, and perhaps the most important. It is also the most frequently neglected. Feedback can tell the sender (i) that the message has been received, (ii) that it has been (or will be) comprehended, (iii) that it has been interpreted, and (iv) whether the receiver is ready for the next instalment. Fully positive feedback indicates success in all four of these; it may be signalled by a simple spoken phase, a body movement such as a nod or a written reply. If feedback is negative, the sender knows that (i) and (iii) have taken place, but there is further work to do on (ii); the sender may also have to adjust initial views as a result of negative feedback on (iii), and should give careful consideration to (iv). Examples of negative feedback include stolid silence, shrugs and restlessness, and tangential replies. Should the sender not notice these, or ignore them, the whole communication process is in danger of breaking down. Either way, positive or negative feedback given by the receiver and used correctly by the sender is a productive part of the communication process. It demonstrates just how interactive and reciprocal good communication should be.

5.4 Basic Theories of Communication

The simplest model of communication is based on electronics and incorporates elementary psychological ideas of stimulus and response. Although its central ideas (transmitter → receiver, stimulus → response) are easy to grasp, this elementary model is linear and mechanistic in the extreme, and cannot be applied to much of what happens in practice. It goes very little way towards explaining why or how 'it's straightforward—ask the boss' can elicit the response 'you must be joking', and doesn't even start to address the issues raised in the example of the advertising hoarding.

The first improvement is to acknowledge that communication is circular and reciprocal (figure 5.1). This elementary model is descriptive rather than explanatory, and needs to be redesigned to include personal and social dimensions. To appreciate this, think about the causes behind the expectations and confusions in the hoarding case, which show that message and context are inseparable.

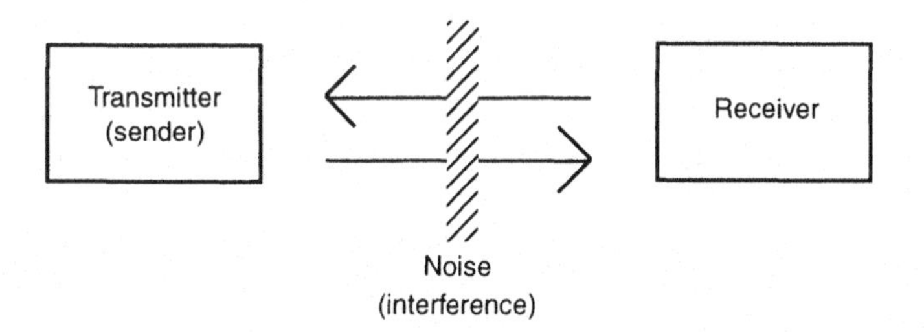

Figure 5.1: *The Bullet Theory of communication.*

Although a contextualised approach to communication is now state-of-the-art, communication theorists themselves were slow to reject the simple mechanistic approach. *Bullet Theory*, as it was called, dominated the first half of the twentieth century:

stimulus→response gun→target transmitter→receiver.

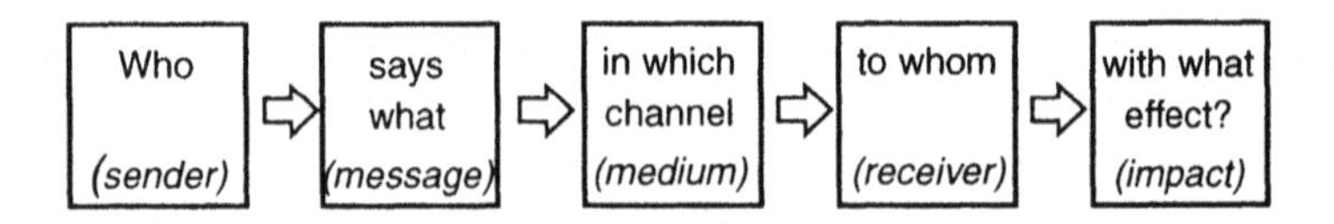

Figure 5.2: *The communication process (Lasswell Formula).*

To see how sophisticated this theory became, we can look at the *Lasswell Formula.* (Lasswell worked with propaganda and examined its effects in both Europe and the US.) In 1948 he began one of his articles with what was to become the most famous single sentence in communication research:

> A convenient way to describe an act of communication is to answer the following: *Who* says *what* in *which channel* to *whom* with *what effect*?

Since then this has become known as the Lasswell Formula (figure 5.2). It is much quoted, but if we probe each of its elements further we find that peeling off a layer simply reveals another. Most of the limitations arise because the theory is linear rather than circular and introduces context only at the very last stage.

After Bullet Theory came *Null Effect Theory.* One of its gurus was Wilbur Schramm. His theoretical models are an advance on bullet theory in being highly circular, as can be seen in figure 5.3. Can we, however, believe that when we communicate we indulge in an activity which will have no effect? On occasions, maybe, but these are the exception rather than the rule. Debate about Null Effect Theory showed that mid-century communications specialists were still underestimating the complexity of the issues, and *Relational Theories* began to be evolved. These incorporate into their models both the linear and the circular elements of the communication process; they deny the idea of any single act of communication being isolated or frozen. An example of relational theory, developed progressively over the last two decades from the work of Dance and Noelle-Neumann, is shown in figure 5.4.

The communication process represented in such models is not simple. It is iterative, with any loop in the central spiral containing

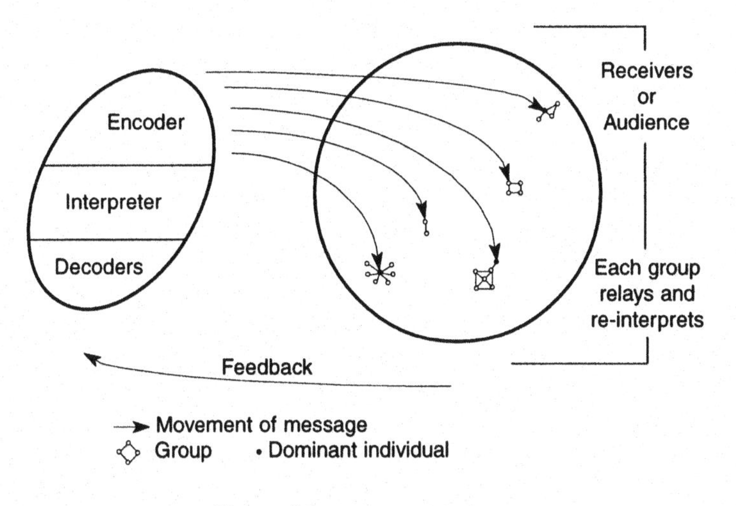

Figure 5.3: *Null Effect Theory.*

all six elements of 'getting the message across', and personal and cultural influences allowed to impinge at any stage. Such models have some parallel with the heredity/environment debate. How much meaning comes from within the person transmitting or receiving? How much results from external conditioning? The models are sophisticated, organic, more generally applicable than any of their predecessors and permit mature consideration of communication issues involving individual perception and social forces.

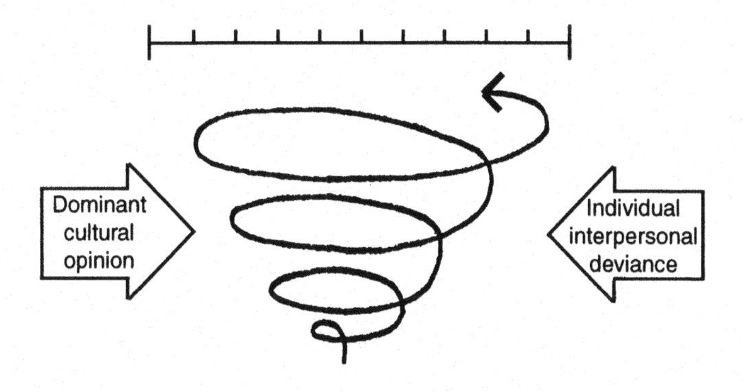

Figure 5.4: *Relational theory of communication.*

If we return to the example of 'It's straightforward—ask the boss', 'You must be joking', we find that these phrases fall within the spiral, itself a moment in time influenced by prior communications and by experience. In this case individual psychological factors predispose each of the participants to espouse or reject the proposed solution, while the dominant cultural climate in which the exchange takes place provides the other controlling dimension.

The importance of either the individual psychological dimension or the cultural sociological one should not be underestimated. In the case of the cigarette advertisement, the two communicating individuals held similar views on smoking and health education; the divergence of interpretation arose from differing cultural background and consequently different expectations. Similar differences occur regularly between communicators of varying sex, rank, educational level, class and interest. Notorious in physics are communications between proponents of nuclear power and environmentalists. As individuals nuclear physicists and Friends of the Earth lobbyists often have different mind sets and the groups within which they work or campaign reinforce those differences. Public opinion on the matter is volatile, swinging against organisations like British Nuclear Fuels when emotive incidents like Chernobyl occur. Relational theories admit all the influences and permit flux.

The end of this chapter has brought us back deliberately to everyday examples like the ones I used at the beginning. In between, we have explored some of the organisational context, terminology, stages and theories of communication. The chapter, and associated study, should reveal that something which we do continuously and tend to assume that we do well, is one of the most sophisticated and subtle processes in existence. All communication, however, shares four features:

- Communication is the creation of more than one person.
- Communication is riddled with convention.
- Mutual understanding requires careful monitoring.
- Social and personal influences complement or counteract any communicator's objective.

Further Reading

Dimbleby R and Burton G 1994 *More than Words* (London: Routledge)

Evans D 1984 *People and Communication* (London: Pitman)

Hargie O, Saunders C and Dickson D 1987 *Social Skills in Interpersonal Communication* (London: Routledge)

Hartley P 1993 *Interpersonal Communication* (London: Routledge)

McQuail D and Windahl S 1986 *Communication Models* (Harlow: Longman)

Noelle-Neumann E 1980 in G Winhoit and H de Bock *Mass Communication Review Yearbook* (London: Sage)

Price S 1996 *Communication Studies* (Harlow: Longman)

Reardon K 1987 *Where Minds Meet* (Belmont, Calif.: Wadsworth)

Torrington D, Weightman J and Johns K 1985 *Management Methods* (London: Institute of Personnel Management)

CHAPTER 6

SPOKEN COMMUNICATION

Shirley Earl

6.1 Introduction

> On his early morning round a Buddhist monk came across two others engaged in debate.
>
> 'It is the flag that is moving,' insisted one.
>
> 'No. It is the wind that is moving,' argued the other.
>
> 'Brothers, brothers, listen to me,' said the new arrival. 'No, it is not the flag that is moving. No, it is not the wind that is moving. Yes, it is your minds that are moving.'

So what is spoken communication? It isn't the vocabulary or the grammar, it is neither inflection nor register, it isn't the person speaking or the person listening, it isn't the non-verbal communication, it isn't the visual aids. Spoken communication is all these things. It is the totality of the shared experience of speaker and audience—oral, aural and visual.

According to John Adair, the contribution of the actual words used in a spoken exchange can be as low as 7 per cent; other vocal features such as stress and intonation contribute another 38 per cent, and non-verbal behaviour such as gesture and facial expression a staggering 55 per cent. In a scientific context the weighting for words and content will be higher, but part of the suffering—at conferences for example—is that presenters pay too much attention to words and too little to the natural world of spoken language. In practice people really do add meaning to their words by using pauses, stress, intonation and non-verbal behaviour.

Essentially there are two dimensions in any spoken exchange—understanding and agreement. In the outcome of the exchange there are four possibilities: understanding with agreement, understanding with disagreement, misunderstanding with agreement, and misunderstanding with disagreement. Look at the four exchanges, A, B, C and D, in figure 6.1, between two Scientific Officers after a meeting with their Head of Division. Each exemplifies one of the combinations of understanding and agreement. Consider the exchanges carefully and place each in the appropriate category. (Not easy, because you are being asked to do something entirely artificial: to consider words quite separately from pause, stress, intonation and non-verbal behaviours. Still ... try it, before you read on.)

What did you decide? I suggest that in A the words imply understanding about the mood of the session which Fred and Jean have just attended and agreement about the future. B involved misunderstanding about the timbre of the session and disagreement about the consequences. In C the participants understand each other's reactions but disagree over the next step. In D both agree that they obtained what they requested materially, but they misunderstand each other's responses.

As these examples show, there can be no guarantee, in spoken or in written communication, that the message which any one individual thinks they relay is the same message their audience receives. There will be a range of correspondence between what is said and what is heard and understood. All the speaker can do is try to maximise this correspondence.

6.2 Spoken Language

In everyday usage we tend to think of a language

- as something specific, like German, Welsh or Cantonese;
- as a tool for naming objects that exist out there;
- as an instrument for expressing thoughts that exist inside the head.

More precisely, a language can be defined as *a body of words (vocabulary) and ways of combining them (grammar) used by a nation, people or race to share meaning*. There is a mystery to

A

Fred	Tough session, eh?
Jean	Just a bit.
Fred	You should have expected it.
Jean	He's not usually aggressive.
Fred	If it involves money he is.
Jean	It involved money.
Fred	Better luck next time.
Jean	I'll be armed!

B

Fred	Tough session, eh?
Jean	Not really.
Fred	You should've expected trouble.
Jean	Why?
Fred	It involves money.
Jean	The Section's within budget.
Fred	He'll have your guts for garters if we overspend now.
Jean	Not mine.

C

Fred	That was a tough session.
Jean	You're not joking!
Fred	Can't have the same next time.
Jean	He's not right you know...
Fred	But you agreed it was tough.
Jean	Because he didn't have the facts...
Fred	I don't want a repeat.
Jean	He'll have to see reason.

D

Fred	Tough session.
Jean	Very.
Fred	I hate that sort of thing.
Jean	I loved every minute.
Fred	We got the money—at a price.
Jean	Not a penny too dear.
Fred	Talk about a pound of flesh!
Jean	What's eating you?

Figure 6.1: *Exchanges representing understanding with agreement, understanding with disagreement, misunderstanding with agreement, and misunderstanding with disagreement. Can you identify which is which?*

language, whether written or spoken: everyone who uses it has an internal conception of the rules which govern it, but no-one can describe everything that can be said by it. Written and spoken language differ, not least because the spoken word is more frequently used, and thus subject to more rapid change and fewer rigid conventions.

6.2.1 Vocal Features

Language, we have seen, is more than just words. It involves other vocal features, such as pause, stress and intonation.

Pause

Pauses in speech are gaps in the sound, a sort of oral punctuation that helps mark out the sense in what we say. Sense can often be clarified by using pauses which are not strictly in line with the grammatical construction of a sentence. Thus a speaker can give a word importance by pausing both before and after it, giving it prominence by isolation. A pause can raise suspense and it is useful for climax or when a strong emotional situation is being described. On the other hand pauses that are too long always break the sense. They can be caused by lapses of memory, nervousness or speech impediment.

Stress

Stress is weight placed to give emphasis. It comes in two forms—word and sentence stress. The stress within words is something we learn along with vocabulary, which is one reason why accents tend to linger. In sentences, nouns and verbs usually receive more stress than adjectives and adverbs. Other parts of speech receive even less. Heavy stressing makes speech laboured and over-emphatic and can be a problem for an inexperienced speaker delivering a paper. *Each Word Is Not Of Equal Earth Shattering Importance.* Nervousness mucks up stress.

Intonation

Intonation is the variation of pitch that helps a speaker convey subtleties of meaning. Speakers' intonation should be closely connected with their thoughts and attitudes; it loses spontaneity when those thoughts lose vitality. Fatigue reveals itself in levelling of intonation, as does incomplete understanding. Intonation can also disclose when a lecturer is ill at ease with the topic.

6.2.2 Non-verbal Behaviour

Beyond these verbal features lies the important non-verbal behaviour that always partners, modifies or augments the spoken word. (Sometimes a speaker's verbal message and non-verbal behaviour seem to be contradictory; it is handy, if not reassuring, to know that on such occasions the audience generally gives more weight to the non-verbal message.) Here are the main features of non-verbal behaviour.

Proximity

This is the interpersonal distance that people maintain when interacting. There are four proximity zones: the public zone (over 4 metres), the social/consultative zone (1.5 to 4 metres), the personal zone (0.5 to 1.5 metres) and the intimate zone (0 to 0.5 metres). Large variations from these norms cause discomfort or disbelief in the words.

Orientation

This refers to both the relative positions of the speakers and to their body alignment. Seating arrangements are genuinely important. The three most common orientations, with their implied meanings, are shown in figure 6.2. Note too that differences in height have implications for perceived status (chairbacks, raised platforms and mortar boards included).

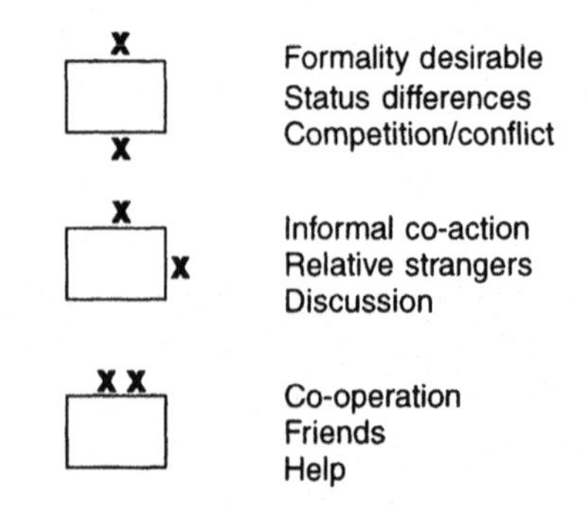

Figure 6.2: *Seating arrangements and their meanings.*

Figure 6.3: *Body language**.

Posture

This conveys information about attitudes, emotions, relative status and motivation. Can you 'read' the body language of figure 6.3?

Touch

This may be used to reassure, console, threaten or defend. Body contact, however, should always be used with caution, particularly with people differing in age, sex or culture from yourself.

Body Movement

Be aware of nods, gestures and movements of the feet in everyday speech, but do not read too much into them. (The body movements

*Interpretation of figure 6.3: A pensive, B joyful, C relaxed, D furtive, E supplicatory, F sad, G celebratory, H shy, I nonchalant, J rejecting.

of people unpractised in oral reporting are truly amazing; if you wish to draw attention to them, be constructive.)

Facial Expression

This is central to spoken communication. One of the main functions of the face is to convey emotional states. Speakers should be aware that they convey meanings by this means and do not always have the movements under control. They can also read their audiences' expressions and modify their message accordingly.

Gaze

This varies from culture to culture. In our society eye contact is only really noticed when it is atypical. Too much or too little is disconcerting. A standard way of riveting someone's attention, therefore, is to gaze at them. Many speakers use this to great effect, knowing that a gaze eight centimetres above the head of a member of the audience appears to be dedicated attention. If the person's attention had been wandering it soon returns.

Appearance

This is a two-edged sword. Speakers can do little about their natural physical characteristics and must initially tolerate the stereotyped and individual perceptions that these physical features engender. In many speaking situations, however, it is important to look the part expected. Attention to dress is appropriate; hair and face can be manipulated.

Paralanguage

Partly pause, stress and intonation, paralanguage also includes interpolation: 'um', 'eh', 'orright', 'er' and all sorts of hisses and splutters. Most of us can become aware of our personal paralanguage and, once aware, be sensitive to and control it. Without these many cues you of course found it hard to categorise Fred and Jean's exchanges at the start of this chapter. The task would have been much easier had you heard their pauses, stresses and intonation and seen their non-verbal behaviours. All are part of spoken language.

6.3 Styles of Speaking

We regularly make use of differing styles, or 'registers', when we speak. Here's an example of one of them, overheard in the departmental corridor:

Dr Mackay	Hello there.
Dr Smythe	Hello.
Dr Mackay	Good weekend?
Dr Smythe	Gardening.
Dr Mackay (nods)	Me too ... the weeds never stop.

Not the sort of stuff that keeps readers of Dick Francis or Catherine Cookson glued to the page, but it does reflect the *casual* register—and displays the sheer banality of many of the real-life exchanges which take place between people every day. To some extent we all take part in these exchanges. They fulfil a useful social function.

As comrades Dr Mackay and Dr Smythe should be trying their best to maintain a friendly but not over-intimate relationship. To ignore each other would be to risk being thought of as rude or, worse, offensive. Their conversation never really took off yet the meaningless pleasantry kept the channels of communication open.

In speaking we regularly use three registers: the *casual* (exemplified above), the *informal* and the *formal*. At the one end lie conversations and discussions marked by false starts and much use of paralanguage. At the other extreme there are the set conventions of formal discourse. Should register, and behaviour, be allowed to become too casual during a group decision-making phase the group will not achieve its objective. As a facilitator in such a group, however, you will soon note the use of casual register and humour to relieve tension.

In daily life we do not consciously search for the appropriate register; rather we recognise inappropriateness. Students learn by watching and listening, by practising, and occasionally by analysis.

6.4 Planning and Presenting a Formal Talk

Talks are of many types. In this section we concentrate on the type likely to be delivered to students or professional bodies and usually referred to as a lecture; other examples are discussed in the following section. This basic guide should suit an inexperienced lecturer or a researcher with a small lecturing remit, but hopefully there will be some points which are new even to experienced speakers.

Never forget that one of the potential strengths of lecturing is that *you* are there in person. Every lecture should convey something about the speaker (preferably something favourable) as well as about the subject.

Giving a talk or lecture has three parts: preparing, performing and reviewing (the 'PPR' pattern).

6.4.1 Preparing

This involves determining the objectives, selecting the content, planning the structure, deciding the method of delivery, and preparing support materials.

Determining Objectives

The impact of a lecture—even its success or failure--depends on whether the audience, and the lecturer, understand the purpose behind it. Are you providing a systematic review of a topic? Are you trying to develop the audience's powers of evaluation? Are you trying to change attitudes? If the objective is not clear to you, you will speak with lack of direction and clarity. A lecture is a formal mode of speaking and it is all too easy to use inappropriate register and phraseology.

Researchers at Stanford have found that regular lecture-goers become expert at spotting imprecisions such as ambiguity and approximation ('type of thing', 'about as much as'), vague numbers ('few', 'some', 'many'), bluffing your way through ('to cut a long story short') and many others. Imprecise expressions cannot be totally eliminated but knowing your objective means you grope less.

Choosing the Content

This may seem straightforward, but even where the topic is clear cut—'the second law of thermodynamics'—care is necessary. Consider the level of the course, likely knowledge of the audience, the time available, and what colleagues will be covering. Most new lecturers try to pack in far too much. Simultaneously listening, understanding, noting and (in the interactive lecture) replying to questions is taxing for members of an audience. And there is the fatigue factor: how many lectures have your students or fellows attended that day? *Your* lecture is not the highlight of *their* week.

Planning the Structure

This involves selecting a method and developing a skeleton. Basic structures include

- lists of topics which are examined in turn;
- hierarchies, where the theme is one of showing relationships between topic groups, topics and sub-topics;
- chains, where the audience is led through argument, proof, derivation, etc, in logical sequence;
- deduction, in which you lead with a general principle or rule and then look at examples or applications;
- induction, leading with specific cases and using them to guide the audience towards a principle;
- problem-based structures, where you outline a problem, discuss possible solutions and finally tempt the audience with a conclusion;
- comparison, in which you take a series of common headings and consider two or more theses under them (matrices are a variation);
- mind maps, where you begin in the middle of a field and build up a linked pattern of concepts or topics.

Deciding the Method of Delivery

This will depend on the nature of the material, the maturity or otherwise of the audience, the size of the audience, the physical surroundings allocated, the range of facilities available and your own personal preferences. Three currently favoured methods are:

- *Talk plus graphics.* Though you may have no visual support for part of the lecture it is inadvisable to talk for 50–60 minutes without some visuals. The pattern of attention in a standard lecture is an initial 20 or so minutes of concentration, a lapse for 10–20 minutes, slight recovery, then renewed relapse continuing to the end. The 'primacy effect' comes into play with a vengeance and gems of knowledge which you relay in the middle, and even at the end, are lost. To keep up interest, oral exposition therefore needs to be varied. Support talk with chalk (writing or drawing), OHP transparencies (legible), slides, videos (with a purpose, not as fillers), models and computer displays.

- *Talk plus demonstration.* Lewis Elton has pointed out that anyone who has seen a Royal Institution lecture will know how impressive lecture demonstrations can be. He himself has spun on a rotating stool and walked on his hands in front of his first year physics class in order to illustrate certain principles. I cannot walk on my hands but can recall vivid demonstrations more than most lecture delivered material. Demonstrations in the lecture, however, should only be precursors to further, possibly laboratory-based, practice, and they should always be tried out in advance.

- *Interactive lectures.* When you distribute handouts you begin to be interactive. Do you *have* to write everything that has to be said on them? Students actually like white space. Most will happily fill in the blanks, complete calculations or invent examples. Once you have let them know what you are doing, and why, they will buzz and snowball, discuss and report back, even in tiered theatres. Work out the logistics in advance however: human necks will not turn through 360 degrees.

Preparing Supporting Materials

If you are using support services, e.g. for duplicating a handout or laser-printing transparencies, allow plenty of time. Book technical support, both personnel and equipment, if they are needed to help with projection, electronic display, or demonstration and, above all, prepare your own *aide memoire*. Notes scribbled in a rush at the last minute are seldom satisfactory. Yet notes may not be necessary; some lecturers work with their pre-prepared transparencies and slides as cues.

6.4.2 Performing

The performance—actually giving the talk—is the second stage of the process. Three important elements of it are orientation, performing style, and variation of audience activity.

Orientation

How many times have you attended a lecture and not known the direction? Flagging (giving a brief overview of directions and sections to be covered) lets members of the audience create their own mind maps for processing what you relay. It is discourteous not to give such a guide, and easy to do so if you are clear about your objectives. If you can't tell them where you are going it may be that you yourself don't know, in which case you are not adequately prepared. Introducing the topic means more than simply announcing a title.

Performing Style

Anyone who has any experience of attending lectures knows how important this is. If the lecturer projects an engaging personality, gives the impression of being interested in the material he or she is delivering and establishes rapport with the audience, the exchange of information becomes enjoyable. Research shows that people who embark on lecturing careers misperceive the process. *Lecturers* say that 20 to 25 per cent of the lectures they themselves attended as students were memorable. *The graduate population as a whole* suggests 4 per cent—a damning figure.

Variety is one of the keys to good performance, and can be applied in many ways:

- Speak slowly when you are making a key point, but use a faster pace for illustrative comment.
- Repeat things when necessary.
- Vary the pitch of your voice. A monotone is a guaranteed soporific. Though the basic pitch of your voice is determined by the length of your vocal cords how you use the rest of your vocal organs is a matter of control and practice.
- Vary the length of your sentences.
- Remember to pause from time to time.
- Use questions and occasional dramatic effects.
- Maintain eye contact with your audience.
- Remember that your hands are mobile and can be used for demonstration and to highlight points. Women naturally make more use of their hands than men. Use, however, should be for a purpose such as emphasis, rather than simply to get rid of nervous energy.
- Minimise reading. The main reason for someone reading is that the speaker is not confident with the material. Reading prevents you establishing rapport and maintaining eye contact.

Variation of Audience Activity

This can be regarded as part of the performance. Timing and planning are the keys. Make use of the periods when attention is likely to lapse and be sensitive to the feedback clues which indicate your audience is ready for a change. Snowballing (pyramiding), quiet time, interactive handouts, instant questionnaires and other approaches all work. The change comes like a breath of fresh air. Think about how to bring your lecture to an end. A good playwright puts poignancy or climax in the last scene; a good actor delivers the final line well. All too frequently lecturers end simply by noting that their time is up (or being informed of this fact by students, audience or Chair). Do not end untidily; do not end on a low; do not simply pick up your notes and leave. Try to summarise the lecture, because such reiteration helps the transfer of information from short term to long term memory. Leave a few minutes for questions, and encourage students to make active use of this time. A question/discussion period helps clear up any problems that have arisen and provides valuable feedback.

6.4.3 Review

Feedback is essential to the review process. Peer review at the point of delivery and student ('customer') evaluation are now established in British universities. What is central to your development as a lecturer, however, is self-monitoring. How did the lecture go? What should you change?

A common review method is to set a test. This can be as simple as 'jot down the three main points of this lecture. I'll let you have two minutes', and then showing the points on a transparency. ('How many of you got all three?') Review tests help to cement learning for the students as well as giving you an instant guide to how much, or little, has been assimilated.

Using a self-appraisal sheet, such as the one provided for Oral Reports later in this chapter, may also be helpful. Self-appraisal sheets can help you monitor your actions, sentence structure, self-image, etc. Used as a matter of routine and filed with your notes, they can be a useful revision/improvement aid for next time.

6.5 Other Contexts for Speaking

We turn now to other situations in which we may be called on to speak. The perspective in each case will be that of the lead planner, a role which only sometimes overlaps with that of lead presenter. Though circumstances change, the PPR routine—preparation, performance and peview—remains essential in every case.

6.5.1 Delivering a Conference Paper

We have all suffered at conferences—from inaudible speakers, speakers who gabble, unnecessary darkness, faulty equipment, unreadable transparencies and a host of other hazards. Yet the recipe for success is straightforward: good conference presenters simply tell us what the presentation is about, why they did the work, how they did the work, what they found, and what they think it means. And then they thank us.

Distilling this out of three months' or three years' research is not easy; conference papers require intense preparation, along the lines laid out for formal talks and lectures (determine objectives, select content, plan structure, decide method of delivery, prepare support materials). Particular attention needs to be given to objectives and to providing structured, meaningful, *correctly timed* content.

Although conference organisers invite 'papers' and will eventually want something that can be published in the proceedings, one of the worst things to do is simply to read out what will eventually appear in printed form. At the actual presentation the full paper need not be conveyed. Some of the best conference presentations—as opposed to conference papers—involve talking round key points or a precis (preferably distributed in advance). If you choose to do this remember to ask the audience to listen, not read, and then direct them to relevant parts of the precis ('If you look at page 2 you'll see the graph ... What surprised us here was...'). Make full use of pauses to let the audience locate the material, and reflect on it.

The recipe for giving a bad conference presentation is also quite straightforward. Bad presenters

- prepare their talk in a hurry;
- believe they know the subject so thoroughly that a logical sequence will emerge naturally;
- don't bother to check the visual aids in advance, or, simply don't have any ('The audience understand how busy I am');
- don't tell people anything about themselves or their co-workers ('They know about me, and co-workers are unimportant');
- make do with a general title;

- forget about an introduction ('Conference-goers are intelligent and don't need flags, they are never tired');
- wear dark and dull clothing;
- fidget if they feel like it;
- keep the lights turned down throughout;
- mumble;
- read long sentences, preferably direct from their forthcoming publications;
- forget about a summary;
- don't look at people;
- over-run ('It avoids time-wasting questions and unnecessary discussion');
- rush away afterwards;
- are too busy to think about what they have done.

6.5.2 Conducting a Seminar

For professional scientists a seminar is an opportunity to disseminate ideas or research by exposition to a group of peers. For students it is an opportunity to scrutinise a topic, present one's conclusions to fellow-students and answer their questions. Seminars are thus peculiar hybrids, with a first phase of formal presentation followed by more open discussion.

For the formal presentation the usual rules apply in the professional context: prepare well, identify the audience, devise visual aids, devise supporting handouts, time the presentation carefully, rehearse and deliver. In the educational context, traditional practice may not be the best training for students. If lecturers ask for seminar 'papers', that is what they get. Papers are read out by students, with minimal interaction, and as a result the ensuing discussion has less chance to take off. If a successful discussion is the aim, a less formal initial presentation may be preferable.

In the professional context the discussion stage is often highly structured, directed procedurally through the Chair. In the educational context it tends to be more fluid (see variations outlined under Discussion in section 6.5.7). Either way the seminar discussion has to be disciplined and effective chairing is essential.

Questions are the key. A good seminar Chair will have one or two questions prepared in case there is an embarrassing silence at the start of the discussion phase (but should not use them if audience response abounds). The Chair needs to watch the audience and encourage short, pertinent questions, ensuring that contributors supply their names and positions along with their questions. Should a contributor be obscure, the Chair's responsibility is to elucidate. Should one contributor try to monopolise the discussion the Chair's responsibility is to silence them—as many people as possible should have the opportunity to ask questions of the presenter. The presenter deserves to be treated fairly, harassment is unacceptable. Chairing a seminar is not easy, nor is presenting the paper and responding to the questions.

Seminar questions come in many forms. They have varying degrees of relevance and range in tone from the supportively friendly to the aggressively hostile. Experienced communicators distinguish the type despite sometimes veiled tone, and it helps if both the Chair and the presenter understand the design and purpose of a question as well as the actual words:

- *Open questions* give the respondent a high degree of freedom. They are broad and frequently require lengthy answers. In a seminar open questions can force presenters to reveal their parameters.
- *Closed questions* do not give presenters any choice. They must answer 'yes' or 'no', choose from imposed alternatives, or simply give a fact. Such closed questions are useful for fact-finding but a respondent who feels pressurised may fail to move beyond the proffered alternatives.
- *Probing questions* are used as a means of encouraging expansion upon points made. They do this by seeking clarification ('Are you saying that...?') and justification ('What evidence...?'). They assess relevance ('How does *x* relate to...?') and can call for further examples and monitor accuracy. Probes may not be easy to answer but they are useful stimuli.

- *Reflective questions* are a technique often used by the speaker's peers. With one pattern of intonation the repetition of something the presenter said simply reveals that close attention was paid; in another tone the same words could be used to focus the collective mood of the audience on a dubious issue.
- *Leading questions* need to be identified and watched. By the way they are worded they guide the respondent in a particular direction, possibly towards false recognition. 'How long ...?' and 'How short ...?' both lead by expectation.
- *Alternative hypotheses,* posed as questions, may prove useful to seminar presenters—the whole process of scrutiny by peers is about exchange of ideas.
- *Multiple questions.* When a hypothesis becomes a multiple question the sequence of points becomes hard to follow for both the respondent and the remainder of the audience. The presenter or chairperson needs to disentangle the parts and ensure each is addressed separately.

For seminars the chairperson and presenter should each remember the 'PPR' approach and

- *prepare* by thinking through the two parts of the procedure, by understanding the nature of questions, by calculating and agreeing timings, and by careful consideration of content;
- *perform* their appropriate role;
- *review* the seminar experience as a whole, and the ideas proffered by peers.

Further advice about discussions is given in section 6.5.7.

6.5.3 Oral Reports on Projects

A project is, in general, a study which at its root is intellectually demanding and concerned with the generation of primary data. These data may be quantitative or qualitative. It may derive from observation in a laboratory or in the world outside, or it may arise from raw data found in libraries or on a field study. A project may be generated by a student, a member of staff or by a commercial or industrial collaborator. It may be carried out by an individual investigator or by a group. Reports on the project can be written, oral or both, and may be presented by individuals, groups or both. Only the oral report given by an individual is dealt with here.

Written and oral reports serve different purposes. A written report may concentrate on such matters as the project's methodology, design, data and interpretation, but the intention of anything spoken is more than simply intellectual. The spoken word is transient and speech is interactive. The *flag* of performance is there, the *wind* of intent is there, the *minds* of the speaker and audience address the shared experience. The moment the oral reporter finishes, however, what was said goes beyond recall. If listeners have not grasped the message there is no way they can recall it. They may question the reporter or ask for points to be repeated but it is unlikely that the same words, intonation or gestures will be used a second time. If readers do not grasp what is written they can reread it. Spoken statements disappear; repetition becomes reinterpretation.

A successful oral reporter will

- *prepare* by seeking guidelines, considering the needs of listeners, allowing for the transience of the spoken word, and by training or practising in a disciplined manner;
- *perform* well by pitching manner and content appropriately, and by harmonising the style of the presentation to the type of data presented (a supervisor or client might well find field photographs extremely helpful, for example, but is unlikely to benefit from viewing slides of storage systems in libraries);
- *review* performance through continuous self-appraisal.

Training and Self-appraisal

Students need training if they are to become skilled at oral reporting. If an oral report will be important at the end of their final year they can be given opportunities for practice in earlier years. This practice can be stepped, for example by concentrating on different components each time. On one occasion you could ask for a report on the application of a technique, on another for a report on problems encountered in data collection, or only for a report of summary findings, or for an explanation of slippage on a project. Training should cover performance technique as well as the selection of appropriate content.

In all cases it is essential that the purpose of the reporting exercise be known and agreed. For example on some occasions a supervisor may be more interested in the method used than the results of the project. And if someone who has not generated the project will be listening to the oral report, more background will have to be given.

Given a modicum of self-discipline beginners and experienced practitioners alike can benefit from regular use of a self-appraisal form, such as the one in figure 6.4. In using it attention should be given to presentation style, the structure of the report and quality of material, its content, and the appropriateness and professionalism of the presentation. The last of these is a holistic measure recognising that in a spoken situation the whole can be more or less than the sum of the parts.

6.5.4 Giving a Public Talk

Public speaking, whether it be at a political rally or a wedding reception, or simply proposing a vote of thanks, can be tackled in a variety of ways. I shall deal with one of the most popular, speaking from prepared notes, before turning briefly to other options: impromptu talks, memorised speeches, and speeches that are written out in full and read. In all cases the personality of the speaker—who he or she actually is—is likely to be more important than in the case of, say, a conference presentation. All public speakers carry multiple responsibilities—to host, audience, fellow speakers and subject.

Speaking from Notes

Whatever the occasion, the talk or speech is likely to consist of

- *an introduction,* to set the tone, stimulate attention and outline the purpose;

- *the main body*, which must have some logical structure;
- *a brief summary*. (Note that in public speeches the summary may be itemised—'firstly', 'secondly', etc. Do not use this technique, however, if the speech is being recorded—you are likely to be edited to 'firstly', 'fourthly'. Note, too, that a sure way to antagonise an audience is to say 'in conclusion' three times in your summary and 'finally' twice and still be on your feet.)

Self-appraisal form for: ORAL REPORT

Topic: ______________________________

Date: __________

Presentation:	(Tick appropriate column) Yes	No
Did I introduce myself?		
Did I introduce the specific topic?		
Did I establish rapport?		
Did I indicate my approach?		
Did I develop the material in an appropriate sequence?		
Did I recapitulate at the end?		
Did I allow myself adequate time?		
Did I invite questions?		

Additional comments:

____

____

____

In a similar situation next time, how might I improve things?

____

____

____

Figure 6.4: *Example of a self-appraisal form.*

Having planned the talk, make a note of key phrases or words on a series of cards (tagged together in order, in case you drop them). Rehearse the talk and modify it, and the cards, as necessary. Memorise the opening and closing sentences if you wish, for safety, but not the rest.

The more self-confidence you can muster the better. Dynamic speakers ooze positive self-image—'I am prepared, I am organised, I look good, You know no more than me, You have invited me for a reason, This is fun...'. They also monitor their performance while they speak—'Am I pacing aimlessly? Gesturing too much? Looking at the audience? Fiddling? Rocking to and fro? Grasping the lectern? Blocking anyone's view?' If you find yourself unable to talk and monitor at the same time, make a checklist for use after the performance.

Applied to this sort of engagement, the PPR technique suggests you should

- *prepare* by ensuring you understand the hosts and the occasion, identifying the audience, choosing the right mood for your speech, selecting content and planning structure, making your notes, and rehearsing;
- *perform* effectively by building rapport, keeping control, being positive, and remembering that your performance ends only when the event ends;
- *review* by monitoring your delivery, during or after the speech, and by continuing self-training if you wish.

Impromptu Talks

Off the cuff talks should only result from last minute requests. Only the most accomplished speakers deliver them well, drawing on a fund of topics and experiences. Beginners may also attempt to talk impromptu, believing that the sudden pumping of adrenalin into the bloodstream will give rise to a scintillating performance. Disaster will result.

But in case you are invited to 'say a few words' impromptu, prepare yourself now. Memorise, and in a few moments of leisure, practise talking your way through, this simple sequence:

1 Think of a message suited to a particular audience.
2 State the message.
3 Put it another way.
4 Give a few examples, or tell a story relating to the message.
5 Restate the message more strongly.
6 Sit down.

The point here is that speaking publicly is not for the amateur. Given a modicum of preparation, however, no-one need be amateur.

Memorised Speeches

These, alas, are also attempted by beginners. They write out their script, commit it to memory, but also carry a sheaf of notes to the podium 'just in case'. The danger is that stage-fright affects memory, and the speaker loses the thread. The way the speech has been prepared means that the speaker is unlikely to pick it up again, resulting in much rustling of paper and embarrassment for all concerned. If you have committed the sin of attempting to deliver this type of speech and your memory fails, take Denis Healey's advice: '*First Law on Holes*—when you're in one, stop digging'.

Reading from a Script

Politicians and scientific spokesman still favour this form of public speaking on occasions when every word may be reported. The inherent dangers are that written words sound pedantic when spoken and the act of reading inhibits both intonation and body language.

The guru for those who believe that public speeches should be written out and then read was Winston Churchill. But he was a leading political figure speaking at moments of national crisis. And he also prepared meticulously. He dictated a first draft without paragraphing, corrected the typescript and adjusted paragraphs, made a recording, played it back for analysis and self-criticism, and rehearsed different stresses and intonation. Only then did he deliver the speech 'live'.

6.5.5 Consultancy

We move now from a situation where one person is 'holding forth', to one where the psychology of *dialogue* becomes of paramount importance. Obtaining consultancy work means shifting the focus of conscious concern from yourself and your discipline to the client. The tactic must be to convert the client's need (in specific cases) or indifference (in general) to active support, funding or collaboration.

In consultancy there is no such thing as an abstract client. One gets into personal dialogues and one cultivates relationships assiduously, from the level of conventional banality, through the conscious neutralising of opposition, to exploitation of commonalities.

Good consultants understand both their own needs and the need of their clients. Are you seeking consultancy work as a challenge? For the money? Because you genuinely believe you have something to offer? Or for self esteem? Or are you seeking it because you've been told to? If the answer is solely the last you are on a losing wicket from the start.

All potential clients need something. As a consultant your job is to identify that need and then create the belief that you can meet it, showing at the same time that you provide value for money. To achieve this you must understand precisely what you have to offer. Is it a report? Or glory, such as publication in joint names? Expertise? A service? Equipment or software? The solution to a problem? Never start down the consultancy route without knowing the answer to this question. The client must understand rapidly what you can provide.

The next step is to make contact. Identify possible clients and offer them something at low cost or free, such as lunch or, more ambitiously, the use of a facility. All the elements of communication should now be in continuous play: words, timing, technicalities, content and non-verbals. For a start meet on neutral territory (fair to both parties or chosen to suit your style) or invite the client to the department (non-verbals and content again—sites visited should impress). Do not be self-effacing. Humility before peers may be suited to seminars but consultancy requires demonstration of expertise by citing and talking about papers, exhibitions and track record. If necessary set up live demonstrations, or invite potential clients to credible events.

Successful consultants must be good at self-promotion and talking. Without talk no client will stay hooked or stay loyal. Obviously the talk needs technical substance but it also needs personal rapport. Having a doctorate or holding a university post is insufficient. Consultants build relationships. They chase, woo and think how to progress; they gain support and agree to collaborate.

Success in one consultancy spawns other projects—in related areas and over a period of time. If you are not invited back or asked to tender next time, that is a measure of failure. Eventually every consultant evolves their own strategy in which what they want and what their clients need are equated.

In summary, the consultant's PPR is like this:

- *Prepare* by knowing why you seek consultancy, understanding what you can deliver, and finding out about potential clients.
- *Perform* by making contact, demonstrating your credibility, cultivating the relationship continuously.
- *Review* by monitoring the progress of the relationship, trying to merge your needs with those of the client.

6.5.6 Sales and Promotion

Sales and promotion is an area that warrants a book to itself, and I can only give some general discussion here.

Selling and marketing are no longer the sole preserve of specialist staff; nowadays organisations expect every employee to contribute to their image. Professionally you may well be forced to sell, propound or proselytise some thing or idea with which you personally disagree. Success lies not in promoting something you believe in but in proselytising something with which you are unhappy or disagree. If you have a negative self-image, change it, because it won't do your organisation any good. Amplify what is good about yourself and monitor how you talk. Winning and holding customers means that everyone should be prepared to accept responsibility.

Effective selling starts with the four questions, who, what, why and how? *Who* are the customers, how many, where? Who amongst them makes the decisions, and in what cycle? *What* do they want and can you provide it? *Why* are you better than rivals? What is your mission—what ideas, products or services do you wish to sell/market? Knowing the customers, why should they buy? *How* can you hook them?

It also involves a choice of style, depending on the relative emphasis you place on the task ('hard sell now') and on cultivating a long-term relationship with the customer. The four main styles can be summarised as:

- *Tell*—high on task and low on relationship (short term return without long term investment).
- *Sell*—high on task and high on relationship (as in the exploration to discover mutual interest advocated earlier under Consultancy).
- *Participate*—low on task but high on relationship (wooing in the long term and being prepared to tolerate an initially low return).
- *Delegate*—low on task and low on relationship but with power of audit and control (as in franchising).

Before you decide which style to use you need to clarify the customer's expectations and be sure you understand them. You need to determine what your company must do in this contract and what is prohibited. You need to know who in the customer's organisation has control of the contract. Constraints must be defined. Quality must be defined. Resources need to be identified.

In the area of sales and promotion, complaining and buckpassing cut no ice. When someone on your staff says, 'Go and see

Accounts', your customer thinks, 'You're too lazy to sort it out'. When you say, 'It's not my responsibility', your customer thinks, 'Then whose is it?' You must produce facts and be assertive as you 'pitch' for business against the competition.

Sales and promotion involves all staff in

- *preparing* by recognising their responsibilities and understanding their customers;
- *performing* well by cultivating a positive image, keeping informed, and using a suitable style;
- *reviewing* by reflecting on corporate identity.

6.5.7 Discussion

Discussion is something which has purpose, is well-managed, facilitates learning and improves the social climate of a course or event. It is quite different from talk, which lacks these attributes (and is what occupies many scheduled time slots in Higher Education and some unsuccessful conference sessions).

The Role of the Leader

As leader of a discussion you need to be clear about its purpose. Which is more important, the *process* of discussion (process-centred), or the *subject matter* (content-centred)? Depending on the balance between the two, you may decide to take an initiating role, providing or eliciting topic expertise, or a listening/ responding role, as prompter, information resource, and reflective observer. In preparing for discussion you need to consider and if possible influence

- the physical setting (low level, comfortable seating arranged in a U or circle has a positive effect on quality);
- the size and composition of the group;
- your own leadership style;
- the objectives and appropriate communication behaviour of the participants.

The most usual way to run a discussion is for the leader to join the group and act as facilitator—by making statements, listening, reflecting and questioning (as in the second of the diagrams below). But there are three other management roles which might be taken:

- *Hands-off.* The manager prepares the task and influences the setting and composition of the group (e.g. by limiting numbers), but then takes a position outside the group, either for the whole discussion or until debriefing is necessary.

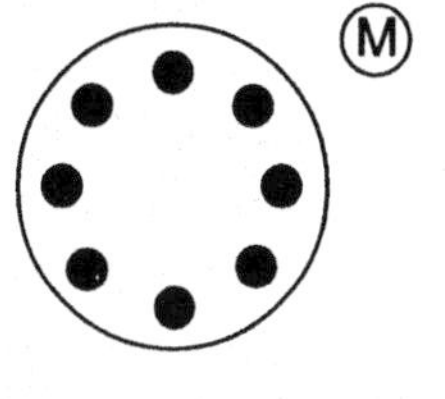

- *Theme Centred Interaction (TCI).* The manager facilitates discussion by making statements only, rather than by questioning. He/she selects themes and encourages participants to consider each theme and their own thoughts and feelings about it. A skilled TCI manager concludes with group thoughts and feelings.

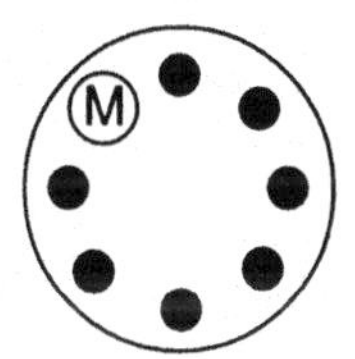

- *Counsellor.* The aim here is personal development through increasing insight, and the manager does not initiate the theme but allows the participants to do so. This can turn into silence or chaos unless the manager marks the boundaries by carefully timed strategies such as the use of reinforcing statements, summaries, skilled listening and non-verbal encouragement.

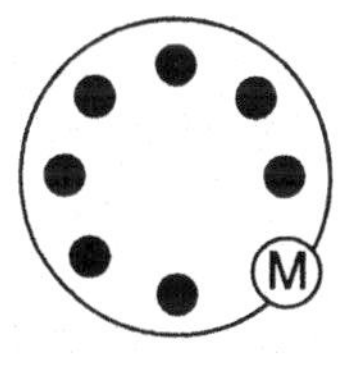

A skilled discussion manager mixes techniques. In the sciences and technology the focus is frequently on content, either to test participants' understanding of the topic or jointly to explore ideas and preconceptions. In the humanities you find more negotiation of meaning and exploration of values or attitudes. In some discussions there will be right answers, in others the outcome must be open-ended.

Participants' contributions can be of many types. They may ask questions, initiate points, express opinions or make proposals; they may support others and agree with them; they may draw others in, build or comment on ideas, or develop separate ideas; they may respond directly or they may block—constructively or destructively. Controlling this too rigidly can defeat the purpose of discussion. But if the group as a whole needs assistance with discussion skills, or even if one or two members are silent, a prompt sheet like the one in figure 6.5 can be used to lay ground rules and improve individual performance.

As leader of a discussion, you should

- *prepare* by knowing your objectives and planning your role;
- *perform* by identifying the boundaries, and by monitoring thinking, behaviour and feeling;
- *review* by evaluating content and experience, training participants when necessary, letting their minds work.

Prompt sheet for: DISCUSSION

	Often	***Some-times***	***Not at all***
1 Did I try to understand?			
2 Did I make suggestions?			
3 Did I support suggestions made by others?			
4 Did I build on other people's suggestions?			
5 Did I agree out loud?			
6 Did I disagree out loud?			
7 Did I probe?			
8 Did I criticise someone else's contribution?			
9 Did I defend my own contribution?			
10 Did I explain or clarify a point?			

Figure 6.5: *Example of a prompt sheet.*

6.5.8 Interviewing

Interviewing is a complex two-way process in which each party constantly influences the behaviour of the other. Nowadays almost everyone is involved in it. Interviews are used for selection or promotion, for briefing or as part of a research project, for counselling, or in the course of appraisal, disciplinary or grievance procedures. Whatever the purpose, interviewing always requires good communication skills and good organisation. These can be taught and learned: productive communication skills like attentiveness and listening, expressing ideas precisely and asking appropriate questions, interpersonal skills like empathy and open-mindedness, and sound organisational skills like preparing, identifying objectives and checking the layout of the room.

To be a good interviewer you must know your objectives and pursue them systematically. Remember that interviewing is a two-way process, in which you give information as well as receive it. You must know yourself; your performance will be under scrutiny from the interviewee, and therefore requires a high level of verbal competence and sensitivity to non-verbal cues.

You must also know the situation. An interview can give you a 'sample' of another person, and you must make that sample as representative as possible. Knowing the interviewee's background can increase the validity of the interview, but only up to a point. Too much pre-knowledge may tempt you to prejudge the interviewee, set the agenda too rigidly or use leading questions.

Preparing to Interview

Start by defining objectives, and list the areas or topics to cover. Four or five is usually enough. Prepare a standard interview form if you plan to use one. It can lead to a more structured approach which pays off in terms of gaining more valid and useful information, and need not result in too rigid and mechanical an interview.

Organise yourself as well as your material. Arrange for privacy and freedom from interruption (including the telephone); consider the seating arrangement (see section 6.2.2). A large desk may be intimidating but easy chairs can make record-keeping awkward. Interviewees are put off if the interviewer constantly shuffles through notes, and if they have to face an untidy desk or the glare from a window. Timing, too, is important; after 40–45 minutes performance declines. If you need longer it is more productive to schedule a second interview.

In the Interview

At all times be aware of possible bias, which may be positive (halo effect) or negative (horn effect). Both can be generated either by the interviewer or the interviewee. As interviewer, however, the responsibility for minimising bias is yours.

Starting tactics are important. Greet interviewees by name, and then make it clear where they are to sit and how long the interview will take. Give an idea of the main areas to be covered; people are more relaxed when they know where they stand.

Signal your attentiveness. The American psychologist Egan recommends a tactic with the mnemonic SOLER:

- Face the interviewee squarely, not with your body angled away.
- Use an open posture. Do not cross arms and legs or otherwise tie yourself in knots.
- Lean slightly forward rather than lounging back.
- Maintain eye contact. Looking away frequently may indicate lack of interest or discomfort.
- Relax. Interviews can be enjoyable, and the preceding four behaviours should also put the interviewer at ease.

Phrase your questions or remarks in concrete rather than abstract terms. Precision in expression avoids ambiguity: 'Can you explain that—*for instance, give me an example*'. And remember to *listen* as well as looking attentive.

Try not to overload the interviewee with information. People have varying capacities, and information that is familiar to you is often novel to the interviewee, who may not be able to cope with what you feel is a reasonable amount of information. Overloading an interviewee is inefficient and may be counterproductive. If you feel you must put over certain information, use visual aids for reinforcement. Don't overwhelm the interviewee with inappropriate language. Avoid jargon and do not over-estimate non-scientists' understanding of scientific issues. It is always possible to check understanding tactfully.

Make brief notes as you go along. Research repeatedly shows that interviewers' recall is patchy and highly selective. Since note-taking can inhibit interviewees and slow down the exchange of information, jot down mini-points and expand them when the interviewee has left. A standardised pro forma can help with data analysis, but qualitative impressions are also useful.

Silence can be valuable, though inexperienced interviewers tend to find it uncomfortable and may be tempted to put words into the mouths of interviewees. Silence gives both parties time to gather their thoughts.

Many interviewers plan the start of their interview but overlook the need to finish effectively. Interviewees should leave with accurate knowledge of what has taken place and a clear idea of what will happen next. At the very least they should be thanked for their time.

Afterwards

Information gathered at interview should be evaluated in a similar fashion to other data: in terms of its validity, completeness and usefulness. Interviewers also need to check the interview process. Were objectives met, wholly or partly? Was the interview cost/time effective? What was the quality of the interview experience?

In summary, the successful interviewer

- *prepares* by researching the situation and background, identifying clear objectives, and organising the surroundings;
- *performs* well by checking the surroundings, starting well, pursuing the interview systematically, using standardised

forms if appropriate, monitoring bias, attending, looking and listening, keeping mini-notes, using silence and finishing effectively;

- *reviews* by checking understanding, analysing quantitative data, evaluating qualitative elements, and checking that objectives were met.

6.6 Using Visual Aids

Visual aids have a number of virtues. They can attract the eye and focus the mind, reinforce the verbal message, stimulate interest, and illustrate concepts and detail which cannot be explained easily or swiftly in other ways. But there is an art to selecting and using them. Visual aids should not be used solely in an attempt to impress and never deliberately to avoid or replace interaction.

These days presenters can choose from a wide range of materials, non-projected, projected and video. Non-projected materials include traditional handouts (verbal, tabular and graphic) and assorted instruction, role and data sheets. The term also embraces chalkboards, markerboards and flipcharts. Projected materials include many of the most useful aids, like overhead projector transparencies, 35mm slides and videotape. Computer-generated displays and multimedia are also available.

Guidelines for making effective use of visual aids are:

- Position equipment carefully, so as not to detract from your presentation.

- Tailor your visual aids to each occasion. If a transparency is dated, the audience will register it; developing outlines as you go along keeps minds active.

- KISS—keep it short and simple. Don't overload pages, transparencies or slides with figures, words or lines, and remember to use white space. A transparency which is a photocopy of a standard printed page will be illegible.

- Concentrate on the audience, not the visual aids. If you watch good teachers using chalkboards you will see that they turn sideways when writing and ask questions as they write, never turning their backs for more than a few seconds.
- Use pointers sparingly. They are for quick identification and not to be played with when not in use.

Variety of visual stimulation reinforces. Discrepancies in standards, however, will show. When you use a graphics department don't let the experts sacrifice clarity to artistic impression. Beware of visual humour; humour is generally better presented orally, if only because you can turn it aside faster if it fails. Don't let over-prepared images, cued scripts or dramatic lighting diminish what you have to say.

The honest solution is to select the medium with which you as an individual feel at ease, and then exploit that medium. If you are artistic with the chalkboard, use it. If you like interaction and video clips, make careful selections. Flipcharts are unlikely to go out of fashion; use table-top versions to flag key words in seminars, and free-standing versions to register discussion points or generate a sequential record by tearing the sheets off and pinning them up.

If you favour overhead projection most organisations are equipped with projectors, and portable versions are not expensive. Use laser print (for clarity) or hand print in medium or broad tips (for individuality), check the focus, and make the changes between transparencies slick. Masking and progressive revelation of a transparency can be fun. If you favour electronic projection, follow the basic rules on text, colour and graphics.

Accomplished speakers learn to control their audience, to direct its attention where they want it. The tricks are simple: switch off an overhead projector when it is not needed, clean the chalkboard between explanations, demonstrate objects only when they are being referred to, avoid plunging the audience into darkness and do not hand items or notes out for people to circulate at random. Point your words, your illustrations, your body and their minds in the same direction.

Further Reading

There are many publications in this sphere. Amongst the more readable and useful are:

Bligh D 1986 *Teach Thinking by Discussion* (Windsor: NFER/ Nelson)

Both V 1985 *Communication in Science: Writing & Speaking* (Cambridge: Cambridge University Press)

Breakwell G 1990 *Interviewing* (London: BPS Books/Routledge)

Clarke A 1992 *The Principles of Screen Design for Computer-Based Learning Materials* (London: Employment Department)

Francis E 1989 *Group Discussion, Learning and Teaching Approaches*, vol.1 (Aberdeen: Robert Gordon's Institute of Technology)

Hargie O, Saunders C and Dickson D 1989 *Social Skills in Interpersonal Communication* (London: Routledge)

Janner G 1990 *Pitching For Business* (London: Business Books Ltd)

Janner G 1991 *Janner's Complete Speechmaker* (London: Business Books Ltd)

Kirkman J 1994 *Guidelines for Giving Effective Presentations* (Marlborough: Ramsbury)

Mandel S 1987 *Effective Presentation Skills* (London: Kogan Page)

CHAPTER 7

Shirley Earl

7.1 Introduction

Meetings are unavoidable—and probably essential—in the modern world. People may meet formally, as on a committee, their meetings may be less formally organised, as when working on related projects in adjacent offices, or they may meet simply because of some spontaneous need.

This chapter is about functioning effectively in various sorts of groups and teams (the word 'group' refers to an *informal clustering* of more than one individual, while 'team' denotes more than one individual involved in *goal-oriented joint action*). We look first at the dynamics of groups, then at individual roles within them, at the concept of leadership, and finally at a number of everyday contexts for group or team work. Two principles underlie the discussion. The first is that we accept the *relational model* of communication—that, within imposed social or organisational limits, people together, by word and action, shape each other's perceptions and reactions. The second is that mutual understanding is taken as being desirable.

7.2 Group Dynamics

Torrington *et al* (1985) have suggested that there are three functional levels of organisational formality[1]:

- *The committee,* which uses structured meetings and has collective power to make decisions. Each person has a stake in the matters being considered and in the collective recommendations. Each member also has a vested interest in the decision and must therefore take a share of the responsibility.

- *People in adjacent offices who need to keep each other informed about what is happening.* They may not take decisions together formally, but personal and electronic networking and the supply of linking information are essential.
- *People who choose to meet to bounce ideas*, work out general strategies, make rough decisions or accept, reject or modify theory before action is taken.

Every scientist has experience of all these—organisational committees, link-pin sessions and think-tanks, be they national conferences or local buzz groups. Other employees' experiences will be similar in kind. An institution's Public Relations Executive may attend more brainstorm and buzz sessions, its researchers more link sessions and its Chief Scientific Officers more committees but the generic types of situation overlap.

In principle the functions and processes of such groups should be beneficial because

- several people can make more time or effort available than one;
- more skill and information shape the decisions;
- a greater variety of ideas can be generated;
- errors can be detected more readily;
- participation increases commitment.

In real life, however, groups rarely realise all these advantages. Working discussions and meetings are likely to be ineffective or even destructive as a result of mishandling information or because of behavioural difficulties in achieving consensus. Studies repeatedly report poor performance by groups for a host of reasons, such as little co-ordination, unequal participation, harm caused by status, power, cliques and sub-groups, pressures to conform or to polarise, the absence of a systematic approach, premature evaluation, changing or ambiguous decision-making procedures.

7.2.1 Reaching Decisions

The way informal or formal groups reach decisions has been studied in much detail. One expert (Bales) has identified three phases of interaction: the *orientation* phase, the *evaluation* phase and the *control* phase[2]. Information is first asked for and given, then the information is analysed, and finally action is undertaken. Bales found that working groups need some members who keep things moving in a co-ordinated way towards a goal and others who ensure that relationships remain harmonious. (See section 7.3 for information about team roles and leading a group or team.)

Tuckman took the analysis further by identifying four types of behaviour in the group as it progresses towards a decision[3]. A group, he said, moves from diversity to cohesion and performance, through stages which he called *forming*, *storming*, *norming*, and *performing*. Each stage involves identifiable processes and has a distinct outcome, but they may blend into a continuous and gradual change in the patterns of interaction within the group. Groups need not always progress towards their decisions in a sequential, predictable fashion.

In the *forming* stage members of groups are likely to be anxious and consequently dependent on any leader. They will be wanting to find out what they can about the situation, task, parameters, permissible methods and standards of behaviour expected. When they discover what these are, resistance frequently surfaces, perhaps in the form of challenges to the leader or in conflict between members. It is at this *storming* stage that opinions become polarised and skilled negotiation may be necessary. An effective group will push through this stage to *norming*, when group cohesion develops and agreed standards emerge. This is characterised by fairly open exchange of views and feelings and by co-operation shown in mutual support. Participants now have a sense of identity and feel they are making progress. Solutions begin to emerge with interpersonal differences being resolved and relationships shifting to ones that help get things done—*performing*. Additional energy seems to be available as final decisions or the completion of the task move closer.

Tensions and conflicts inevitably arise and there is ongoing debate about whether these can or should be resolved. The *consensus* perspective emphasises agreement, co-operation and stability—scientists are members of a community, fraternity, the same team, department or organisation and as such should subscribe to similar values and share roughly the same interests. The *conflict* perspective challenges the idea of underlying harmony. It suggests that different groups arise because of mutually opposed experiences and objectives. Unity is only achieved when a dominant group or faction manages to represent its specific interests as universal. Cohesiveness is therefore relative—nuclear physicists recognise the benefits of accelerator research; CERN and Daresbury may become factions. But for any scientific organisation to function there must be some degree of cohesiveness, which generates pressure towards agreement.

So long as risks are controlled, however, the constructive value of conflict should not be underrated. Priestley could have done with advice from Lavoisier whose schemata had room for the analysis of air into gaseous elements, but late eighteenth-century scientific problem-solving was a competitively individualistic affair and there was nobody around to warn Priestley that his schemata were unsound. His case illustrates the point that there can be all kinds of holes and flaws in one individual's representation. Shared representations—schemata that are collectively explored—are more likely to provide a sound basis for problem-solving.

Research into decision-making has yielded some encouraging results. Hall and Watson, for example, found that the decisions produced by groups were of better quality than those produced by the individual members alone, and that groups given instructions on how to work towards consensus produced better quality decisions than groups working without guidelines[4]. Hall and Watson's guidelines have since been extended and the standard recommendations for successful group decision-making are:

- Avoid arguing for your own opinion, instead present your position lucidly and clearly.
- Avoid 'win–lose' statements, indeed, try to discard any notion that someone must win and someone must lose.

- Avoid changing your mind only to reach agreement and to dodge conflict.
- Avoid conflict—conflict-reducing techniques like majority votes and trading are available.
- View initial agreement as suspect, explore the underlying reasons.
- View differences of opinion as natural and helpful.

7.2.2 Groupthink

Talking things through in a group is a good way to develop sound representations but it has its dangers, the most prominent being *groupthink.*

If the desire of a problem-solving group to reach agreement, or the pressure from a directive leader, is too strong, then poor quality solutions will result. Managers are taught to beware of the phenomenon, which arises from a mixture of high cohesiveness, directive leadership, stress, belief in limited alternatives, group isolation and insufficient attention to method.

Symptoms of groupthink include the idea that the particular problem-solving group is superior and unassailable and occupies high moral ground, stereotyped and shallow depiction of people and ideas outside the group, and direct pressure on dissenters. Individual members of the group are likely to use self-censorship, some may become self-appointed mindguards who protect the group from criticism, and an illusion of unanimity is created.

Groupthink lowers creativity and can undercut the benefits of shared decision-making. It does so because limited information is handled selectively and objectives are not fully surveyed. Alternatives aren't appraised, the risks of the proposed solution are not examined in full and neither a fall-back position nor a contingency plan is worked out. The theory of groupthink was applied by its creator (Janis) to the Bay of Pigs fiasco, in which 1400 Cuban exiles were meant to overthrow the entire Fidel Castro regime, but it applies equally in other spheres[5].

7.3 Group and Team Roles

7.3.1 The Effective Team

A group which shows signs of groupthink is a sick one. An effective working team acts in sharp contrast. Optimally it has between four and ten members who share a common understanding of purpose. They regularly review their progress and performance against the objectives and timescale and operate under agreed leadership rather than through formal control from the Chair.

Field surveys by The Industrial Society and other bodies show that effective teams use recognised problem-solving techniques (such as a Past/Present/Future Approach, Delphi Technique, or Quality Circles). Such teams also encourage new ideas. Their members practice active listening and each person takes responsibility for individual tasks. Members feel they have collective ownership of the team's work and are prepared both to give and to receive help. Decisions, actions and progress are recorded succinctly. To achieve this there must be five-way management within the team: up, down, across, out (to other teams, customers and suppliers) and, most important, self-management. Each member knows the key areas in which his or her personal contribution is required. Members are able to prioritise, can set realistic targets and can monitor their own performances, applying self-knowledge as well as technical knowledge. Training techniques and career reviews now foster this. According to Belbin, effective teams need[6]

- *a Chair/leader*, who, as necessary, appreciates what the objectives are, makes sure all views are heard and keeps things moving;
- *a company worker*, who is diligent and keeps the organisation's interests to the fore;
- *a completer/finisher*, who follows things through and suggests conclusions;
- *an ideas person*, who has imagination and can make novel suggestions;

- *a monitor/evaluator*, who has a sound sense of judgement and can assess contributions and progress;
- *a resource investigator*, who has the ability to make contacts, can respond to a challenge and is able to weigh up whether things are feasible;
- *a shaper*, who has drive and can influence by follow-up;
- *a team worker*, who promotes team spirit and responds to changes in the situation.

Most people are capable of fulfilling more than one of these roles so it is handy to check the roles members are fulfilling in any team with which you work. Should any functions not be being met members can adjust their performance to suit. Team members should be encouraged to identify their own current strengths and weaknesses.

7.3.2 Leading a Group or Team

There are three notable schools of thought about leadership—the *trait* school, the *situational* school and the *need* school.

The original approach was to consider that leaders are born, not made. Hundreds of research studies were devoted to consideration of physical characteristics like height and appearance, to ability characteristics like intelligence and fluency in speech, and to personality *traits* like extroversion and dominance. There was little correlation between findings and no agreement about characteristics. In consequence theorists now favour *situational theory*—leadership being contingent upon demands of the circumstances, or *need theory*—leadership arising out of the joint needs for completion of task and maintenance of social relationships.

Experiments indicate that democratic leadership is more effective than either autocratic or *laissez-faire*. In a classic study groups of children were asked to carve models from bars of soap, each group being exposed to a different style of leadership—authoritarian, democratic or *laissez-faire*. Authoritarian leaders did not participate in the activity, made decisions and assigned tasks without consultation. Democratic leaders were friendly towards group members, participated in the activity, sought maximum involvement and gave reasons for praise or criticism. *Laissez-faire* leaders were passive, made neither positive nor negative evaluations and did not attempt direction or co-ordination. The democratic style was found to produce the highest morale and high quality models; in the absence of the leader, members kept working. The authoritarian style produced most models but when the leader was absent members' production declined and behaviour deteriorated. *Laissez-faire* produced fewest models and much disruptive behaviour.

Subsequent research with both adults and children has largely confirmed this pattern. Note, however, that psychological research may be culturally biased towards the democratic style and that there is nothing to prevent any individual leader mixing their behaviour, i.e. exemplifying all three styles at different points of time during one group session or different styles at different sessions.

According to situation or need theories of leadership, any individual member of a group has the potential to lead, but the likelihood of this occurring will depend on a range of factors. These relate to the situation (the tasks which the group faces and their level of difficulty, the physical setting in which the group operates and the composition and history of the group), to the group (the expectations of the corporate body, the relationships between members and the need or not for innovation) and to individual characteristics (whether a person's particular style suits, and the acceptability of individual ways of achieving goals). Given the freedom each group selects or recognises a leader, the selection depending on members' perception of, for example, rank or expertise as the basis for authority, the importance of maintaining harmonious relationships, and the personalities of other members. The traditional isolated and aloof leader, in the position as a result of birth or privilege, may still survive but is not chosen naturally and is no longer culturally favoured. The person who exemplifies leadership (perhaps different from the person who is designated leader by an outside agency) is someone who is able to see and exploit a whole situation. Such people are aware of, and are sensitive to, things going on in the group around them. The difficulty is that any one leader only knows what he or she personally experiences. To optimise and retain their positior leaders must therefore cultivate conditions which enable all members of the group to use their intelligence and skills, to share goals and to work together. Leaders must stay alert to changes. Indeed, 'When the best leader's work is done the people say "we did it ourselves".'

Management theorists currently expect leaders to

- be learners not masters;
- possess self-esteem and self-confidence;
- question and listen;
- value people;
- share;
- build relationships;
- have vision;
- both develop and implement action;
- model the way;
- be able to discipline; but
- be positive and supportive rather than complaining and blaming and deal effectively with criticism.

The approach is called *Action Centred Leadership* (ACL). It presents a model of the leader in orbit around three interlocked circles (figure 7.1). The inward pointing arrow represents the leader taking an overview of task, team and individual, while the outward pointing arrow represents the need for the leader to manage the interface between their own team or department and others either inside or outside the organisation. The espousal of ACL explains why, in training people for leadership, strong emphasis is now put on interpersonal skills: information-giving, information-seeking, directing and delegating, co-ordination, evaluation, encouraging, harmonising, relieving tension and self-appraisal.

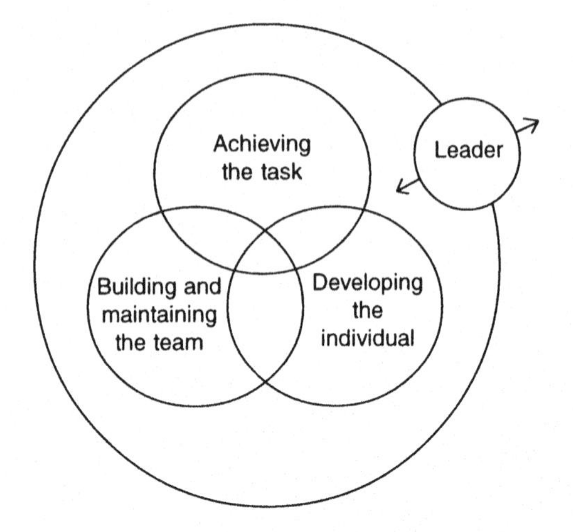

Figure 7.1: *Action Centred Leadership.*

7.4 Contexts for Group and Team Work

7.4.1 Working Groups

Working groups are composed of specialists (either voluntary or nominated). They meet over a finite period and are empowered, fully trusted, with the responsibility and authority to carry out a particular job. All members (or role-playing learners) therefore need to know the nature and scope of the task involved, the nature and extent of authority vested in them, the time scale involved, the method to be used, and the results that need to be achieved.

Working groups should not be told what to do, since the combined experience and knowledge of the members should be sufficient for them to obtain the necessary results. If members are overloaded or fear criticism, the process breaks down.

7.4.2 Committees

Committees are elected or appointed bodies which have clearly defined, continuing organisational remits to tackle matters too complex or demanding for individuals to handle alone. The process by which committees function is based on the convention of formal meetings and agendas, demanding the chairing skills of the leader and active participation on the part of members. Committee procedure is an awkward thing to teach—tedious if covered didactically and full of pitfalls for the unwary. Don't attempt it if you're not thoroughly familiar with, and preferably experienced in, forms and custom. It is simpler and less stressful to outline only the basics and let simulation or practice reveal technicalities, variations and minefields.

The foundation structure is the set agenda which begins with a record of attendance and apologies, proceeds to minutes from the last meeting and then goes through named items. Reports and matters like financial statements are placed before topics for open discussion, and the agenda ends with any other (relevant) business. Certain customs, all aiming for efficiency, surround the use of agenda. A week before any meeting takes place it is customary to send copies of the agenda to everyone entitled to attend—time is money and if no matters that seem to concern an individual are scheduled he or she can choose to send apologies and devote the time to other things. The agenda for a committee meeting is usually prepared by the secretary in consultation with the chairperson.

Before the meeting the *person taking the Chair* should

- ensure that everyone knows when and where the meeting is to take place;
- check that actions allocated at the last meeting have been carried out;
- brief any specialists needed for the meeting;
- ensure everyone knows what is to be discussed;
- ensure that the committee room is set up satisfactorily;
- mentally allocate time to each item, so that the most important ones get most time.

During the meeting the Chair should

- start on time;
- have scheduled breaks if the meeting is long;

- select appropriate procedures for discussion;
- give regular feedback on progress, what has been agreed and what has still to be discussed;
- plan the implementation of any decisions, including allocating actions to individuals;
- review and check the record of discussion and decisions;
- make sure everyone is involved;
- not fly personal kites;
- permit the expression of unpopular opinions but not of person animosities;
- finish the meeting on time.

Committee Chairs monitor the function of the team and facilitate its work. They should progress, listen and watch rather than make decisions themselves. They should also aim to minimise the often subconscious attempts of strong committee members to beat weaker ones.

After the meeting the Chair should

- check the accuracy of the record of the meeting;
- make sure that everyone entitled to attend receives a copy of the record;
- make sure that anyone charged with action knows what to do;
- check that copies of (or extracts from) the record are circulated to other appropriate people or committees.

Before a committee meeting *members* should

- know what the meeting is for;
- check that any personal action arising from the last meeting has been taken;
- read advance papers;
- make pertinent notes;
- identify their overall role.

During the meeting members need to

- work through the Chair, recognising the authority of that role;
- use social skills to interact and persuade;
- avoid personal attacks on others;
- support, develop, modify or, if necessary, present the case against contributions from others;
- be objective in seeking solutions.

After a meeting members should

- review their participation;
- advise anyone they represent of decisions;
- consult anyone they represent about trends;
- study, and if necessary correct, the record of the meeting;
- take any assigned action.

7.4.3 Syndicate Sessions

At the time of writing, universities, research institutes and the Civil Service still make wide use of working group and committee systems. In contrast, private businesses and multinational companies are moving towards team building, committee-free systems, the use of task forces and the use of *syndicate sessions*, sometimes assisted by professional facilitators.

To work in a syndicate you need to have mastered the art of *active listening.* Listening is a fundamental but under-researched phenomenon. You learn to listen before you learn to speak, you learn to speak before you learn to read and you learn to read before you learn to write. On this scale listening is part of the foundation of all communication, carried out by your brain as well as your ears. Listening involves hearing and seeing, searching, selecting, assimilating, organising, retaining and responding both verbally and non-verbally. It is fundamental to interaction. Hargie argues that there is passive, covert listening and there is active, overt listening[7]. Passive listening omits response and can disorient or mislead other people in group situations. Active listening occurs when you display behaviour which indicates that you are assimilating and reacting.

Consider a training morning on team-building and problem-solving, whose programme includes:

1 Focused talks, followed by facilitated identification of problems, for the whole group.

2 Group problem-solving (one issue per group), done in syndicates.

3 Facilitated reporting back and debriefing, for the whole group.

Active listening would be of paramount importance for all participants but emphasis on the type of listening would vary. Thus the focused talks call for comprehension listening—the emphasis being on participants searching for, assimilating and retaining the facts and the theme. The identification phase calls for evaluative listening—when participants relay responses which indicate selection and the amount of assimilation so far. The syndicate stage calls for further comprehension and evaluation plus appreciative and empathic listening—participants seek out and become involved with further messages and, by word and gesture, physically demonstrate their responses to each other. The facilitated reports call for additional listening of all types, but debriefing reverts to emphasis on comprehension.

The main obstacles to active listening during syndicate sessions are noise, attentiveness and individual bias. Noise may be extraneous but it also arises when participants talk across each other and members have to try to assimilate information from more than one source at a time. Inattentiveness hardly needs explaining; it is the absence of concentration but should not be regarded as necessarily ill-intentioned—pondering ramifications is a natural and necessary activity. Syndicate sessions are often pressurised, and a skilled facilitator will always move between groups, watching for the blocking behaviour or avoidance of difficult realities which arise from judgmental bias.

7.4.4 Group Projects

The whole area of group projects and problem-based learning is in a state of flux—see Boud and Feletti (1991) for a useful compendium of struggle, successes and failures[8]. The Employment Department has sponsored the publication (ed G Gibbs) of case-studies where peer, self, tutor and group assessment has been designed to complement problem/project methodology. Modules 5 and 7 of the CVCP Universities' Staff Development and Training Unit compendium on active learning also contribute to the debate[9, 10].

Group projects can provide an opportunity for employees, researchers and students to draw on past experience of interactive situations, as well as on their personal and professional skills. Each project should advance members' knowledge and improve their technical skills. It should also widen their understanding of interaction.

Careful structuring is the key to success. Structure the group problem to link with what the team members already know, and to give them the opportunity to advance. Explain that a group project is active experiential learning, and indicate what you yourself see as the essential process and core skill (be it consensus, delegation and division of labour, or listening). Talk about what has been examined in this chapter—the benefits of working in groups, the problems encountered by groups, the stages of interaction, and the nature of leadership. Consider the composition of the group as well as the relevant scientific theory and applications.

Some Engineering courses now use cross-year teams, deliberately reserving roles as leaders and resource investigators for senior students and using junior students only as team workers. Some Mathematics courses have been stepped to simultaneously advance modelling expertise and communicative competence. Determined tutors who practise Action Centred Leadership themselves can locate commercial and industrial partners willing to provide both ideas and resources for group projects.

References

1 Torrington D, Weightman J and Johns K 1985 *Management Methods* (London: Institute of Personnel Management)

2 Bales R and Strodtbeck F 1951 'Phases in group solving' *Journal of Abnormal and Social Psychology* **46** 485–495

3 Tuckman B 1965 'Development of sequences in small groups' *Psychological Bulletin* **63** 384–389

4 Hall J and Watson W 1970 'The effects of normative intervention on group decision-making performance' *Human Relations* **4** 299–317

5 Janis I 1972 *Victims of Groupthink* (Boston: Houghton Mifflin)

6 Belbin R 1981 *Understanding Organisations*, edited by C Handy (Harmondsworth: Penguin)

7 Hargie O, Saunders C and Dickson D 1987 *Social Skills in Interpersonal Communication* (London: Routledge)

8 Boud D and Feletti G 1991 *The Challenge of Problem Based Learning* (London: Kogan Page)

9 Griffiths S and Partington P 1992 'Module 5: Enabling active learning in small groups' *Effective Learning and Teaching in Higher Education* (Sheffield: CVCP/USDTU)

10 Horobin R, Williams M and Anderson B 1992 'Module 7: Active learning in field work and project work' *Effective Learning and Teaching in Higher Education* (Sheffield: CVCP/USDTU)

Further reading

Adair J 1989 *Action Centred Leader* and *Effective Communicator* (London: The Industrial Society)

Adair J 1990 *Not Bosses but Leaders* (London: Kogan Page)

Argyle M 1983 *The Psychology of Interpersonal Behaviour* (Harmondsworth: Penguin)

Hartley P 1997 *Group Communication* (London: Routledge)

Kolb D 1984 *Experiential Learning—Experience as the Source of Learning and Development* (New York: Prentice Hall)

Pennington D 1986 *Essential Social Psychology* (Sevenoaks: Edward Arnold)

INDEX

Lightning Source UK Ltd.
Milton Keynes UK
UKOW05f1654221116
288295UK00024B/618/P